每家必備的煮餸秘笈

女人必學100道菜

蕭秀香(三姐)・江美儀 合著

萬里機構

推薦序一

早年我曾邀請三姐上我的電台節目《金漆招牌》，讓我有機會更深入了解這位中菜女廚。我認為三姐是個有急才、而且應變能力高的人。烹飪從來不是靠死記硬背、不是照板煮碗就能煮出一碟好餸，三姐的獨特之處，就是能因應不同食材和客人口味而烹調出合適的餸菜，能夠就地取材，煮出用心之作，令人津津樂道。

三姐有句說話我印象特別深刻，就是「原來唔識，是不可以靠搏的」，其實這句話可應用於任何情況，所有事都非靠僥倖，就正如三姐的廚藝並不是一朝一夕練成，不靠走捷徑，而是靠穩打穩紮，日復日、年復年浸淫出來的。這本書是三姐多年來的心得，絕對是有意學習烹調的讀者最佳入門手冊。

《女人必學 100 道菜》，跟三姐學烹調何止 100 道菜呢？我期待三姐快些有第 200、300 道菜，與大眾分享！

張宇人

香港立法會（飲食界）議員

推薦序二

籌備開拍《女人必學 100 道菜》，因為飲食節目是最受觀眾喜愛的節目類型之一，而疫情以來，市民減少外出賦閒在家，入廚炮製美食不只是排遣煩悶、慰藉心靈良方，亦是為觀眾打打氣的絕佳方法。

一個廚房，兩個女人，絕對是節目的全新嘗試，而三姐及江美儀一直是製作團隊心儀的主持人選。

三姐廚藝超群，菜式不拘泥於傳統，力求多變，擅長將酒家複雜菜式，變成家中餐桌美食，多年來擔任《流行都市》節目的廚藝專家，她的菜式已是一眾粉絲們心目中的必學菜。

美儀熱愛美食，廚藝令人驚喜。為令觀眾有所得着，每次入錄影廠前，她必先反覆試煮，務求每道菜式由選材、配搭、煮法、擺盤，皆力臻完美。此專業態度，正正是她多年來深得觀眾喜愛的原因。

25 集節目，兩位主持火花四濺，笑語連珠。台前幕後 100 分付出，製作 100 道菜式，希望觀眾看得開心之餘、繼續期待驚喜、享受美食，樂觀面對每一天。

張志明

《女人必學 100 道菜》監製

推薦序三

最初「認識」江美儀，是在溫哥華看到劇集中她的演出，令人眼前一亮！

2000 年返港拍《廟街・媽・兄弟》，原來她是我樓上的屋主！

每次家中宴客，都會請她一齊來吹吹水！

這小妮子能言善道，而且對烹飪顯得蠻有興趣，每當我講到烹調的秘訣時，她都默默記住，每過一段日子，她都會歡天喜地的告訴我，成功做了你的豉油雞、蒸水蛋——不意外，本來就是入得廚房，出得廳堂的好苗子！

今天她出烹飪書了，但我尚未有機會品嘗她的手藝哩！江山美人儀，妳欠我的一餐，幾時還？！

黃淑儀

推薦序四

記得《女人必學 100 道菜》其中一段對話是這樣，美儀問：「三姐，妳今日嘅材料，樣樣九唔搭八，風馬牛不相及，想點呀？」三姐答：「必會令妳意想不到！」就是這樣的一問一答，充分展現了超級女廚神和超級女演員的神級氣度。

一對寶貝，煮餐飯都嘻嘻哈哈，開懷滿足，輕鬆下廚，令人看得開心愉悦。一餐開心的飯，令大眾領略烹飪的樂趣。

粵式菜系，既可大排筵席，亦可變化萬千為粵式小炒、家常小菜，正如三姐所言，意想不到！

三姐經常話要慳儉，如何將一些頭頭尾尾食材，不浪費，以大廚功架，點化成小家名菜，你不喜歡的食材，她也弄成百味珍饈。

這就是三姐的烹調哲學。

食在廣州，廚出鳳城。這位鳳城三姐，將順德廚藝玩到盡，既可堅離地，也可勁貼地；人揸鑊鏟，她揮鐵勺！

世代不同，從前看烹飪節目，可能只做一個芥蘭炒牛肉；弄到芥蘭翠綠，牛肉滑香，觀眾就覺得很不得了啊！來到這世代，煎炒煮炸炆燉焗，三姐同你餐餐爆玩；就連喜茹素的美儀，晚晚都同你整個靚餸。

一百道菜，食足一季，不過……要親手下廚弄一餐，就必定要看這本《女人必學 100 道菜》啦！

秋冬已臨，三姐、美儀該為男士做點貢獻，推出男人壯陽 100 道菜，陰陽調和，那就功德無量了！

安德尊

推薦序五

認識美儀已經有一段時間了，從互不相識到成為同事，然後一起跑步，之後再毫不客氣地上她家作食客。記得有一次，我真的以為是一道酸菜魚的菜式，沒想到神來之筆的一個驚喜～原來是素酸菜魚。對於每個月會吃素幾天的我來説，美儀這個食譜真的是我的飲食聖經。要知道，在香港素食的選擇真的沒有台灣豐富，加上我們日夜顛倒的工作時間，有些時候吃素食真的有點難度。現在不用煩惱了，我們可以自己煮。而最重要的是，食譜裏面很多都是快、靚、正的烹調方法，不單止製作簡單、容易上手、味道更加非常美味。

很多人覺得煮得一手好菜可能是一個女人的秘密武器，但對於我來説，這更是善待自己的一個生活調劑。

謝謝美儀，讓我煮得更精彩！

唐詩詠

蕭秀香（三姐）自序

人生經歷起起伏伏，絕大部分時間都在爐火前與男人共事，伴隨煤氣爐和食具，演出一首首廚房交響樂。

《女人必學 100 道菜》帶給我的是截然不同的體驗，一本書能夠出版，這個機會得來不易，我衷心感謝每一位於《女人必學 100 道菜》付出過努力的隊友，多謝 TVB 提供一個非常好的平台，製作團隊非常專業，同時有不少空間給我發揮，亦要多謝美儀這個幕前經驗十足的好拍檔。希望這本書帶給大家不單是 100 道菜，更能激發五味糾纏的催化劑。

烹飪是我的興趣，能將興趣變為我的工作，見到每一個食客，滿面笑容，滿足地離開餐廳，讓我感到安慰。不同食材的配搭可煮出千變萬化的菜式，不斷嘗試，是我對烹飪的執着！

眾所皆知，美儀是一位資深主持人和藝人，起初當然需要磨合，不過於拍攝數天後，經已互相配合得很好，默契十足。一個入得廚房，一個出得廳堂，正正就是最完美配搭。

2005 年參加「食神爭霸戰」，憑着一道「蓮漪飄香」勇奪金獎及最具創意獎。2007 年獲法國國際廚皇美食會頒發藍帶白金五星獎，兼入選中國飯店名人百福榜。最近更以一道「酒香瀨尿蝦」，獲得世界粵菜廚皇三粒星的評價。

現時為香港餐務管理協會副會長、世界粵菜廚皇協會會員、世界中餐名廚交流協會顧問。曾擔任香港旅遊協會飲食評判、澳門工聯會飲食評判。

三姐秉持「傳統有創新，創新不忘本」的理念烹製和創作菜餚。除經營飯店外，三姐還擔任電視台節目廚藝專家，包括《流行都市》、《都市閒情》等，分享入廚心得，廣受歡迎。先後出版《食盡其材慳家菜》及《大廚小菜（增訂版）：三姐的 50 道創意家常菜》。而她在將軍澳開設的食肆在城中頗具名氣，更是很多藝人的飯堂，食客甚至遍及海外。

江美儀自序

我從來沒有想過會主持一個烹飪節目，更無妄想過會出本烹飪書，還要是和專業女廚神三姐合作，這簡直是光宗耀祖的大事！

一向以來，我視烹飪為興趣，偶爾作為消遣活動，既可滿足味覺又可供奉五臟廟，超好玩！但要主持一個節目，就是兩碼子事。幸好我的搭檔是三姐，一個同我性格相近，做事認真又循循善誘的前輩，所以我們合作得非常愉快。

除此之外，多謝一班幕後敬業樂業的工作人員，為我們做好前後期製作過程的準備功夫。至於現場攝影隊友，他們每次爭先恐後去試食，又大家一同爭食雪條的情景，我到現在想起來還是會心微笑！

總括來說，感恩我有這個機會去分享一下我回憶的味道和推廣健康素食，讓大家可以多一個選擇，既煮得開心又食得健康！

作者簡介

香港女演員及主持，演技廣受認同，曾奪萬千星輝頒獎典禮2013「最佳女配角」殊榮。

部分代表角色包括：易懿芳（三姨太）《名媛望族》、方芮嘉（Head 姐）《衝上雲霄 II》、嚴查向善《親親我好媽》、姚淑嫻（Lulu）《殺手》。

幕前角色多變，台下巧手出眾，追求色香味美滿分配搭，喜與良朋好友如 Gigi 姐一同鑽研廚藝，視美食分享為快樂泉源。

目錄 CONTENTS

推薦序

自序

初體驗菜
Beginner level recipes

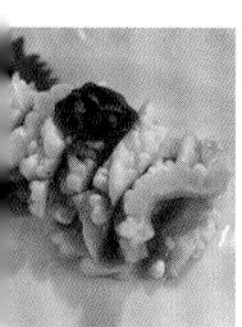

進級挑戰菜
Intermediate level recipes

升級不敗菜
Advanced level recipes

宴客分享菜

Party and banquet favourites

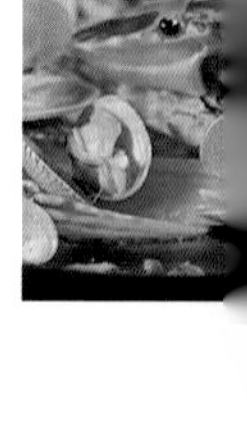

有營素菜
Nutritious vegetarian recipes

三姐、美儀常備配料

三姐、美儀各有看家的烹飪本領，美儀的私房冬菇能將冬菇的香氣提升，無論做主角或配菜皆宜。三姐的甘草欖角惹味香濃，對菜式有畫龍點睛的效果。

美儀私房冬菇

Elena's signature steamed shiitake mushrooms

材料

冬菇 15 朵
薑 2 片（拍扁）
葱段 2 條
生抽 2 茶匙
糖 1 茶匙
紹興酒 1.5 湯匙
蠔油 2 湯匙

做法

❶ 冬菇洗淨、浸軟、去蒂（水留用），加入生粉拌勻，用水沖洗，瀝乾水分。

❷ 以生抽、糖、紹興酒、蠔油醃一會，放蒸碟內，放上薑片、葱段及適量冬菇水，以耐熱保鮮紙封好，以中火隔水蒸 30 分鐘即成。

* 蒸製過的冬菇隨時可用，或存放冰格備用。

Ingredients

- 15 dried shiitake mushrooms
- 2 slices ginger (crushed)
- 2 spring onion (cut into short lengths)
- 2 tsp light soy sauce
- 1 tsp sugar
- 1.5 tbsp Shaoxing wine
- 2 tbsp oyster sauce

Method

❶ Rinse the mushrooms and soak them in water till soft. Drain and set aside the soaking water. Cut off the stems. Add potato starch and mix well. Rinse off the potato starch and squeeze the mushrooms dry.

❷ Marinate the mushrooms in light soy sauce, sugar, Shaoxing wine and oyster sauce. Arrange on a steaming plate. Arrange sliced ginger and spring onion on top. Pour the soaking water over. Cover in microwave safe cling film. Steam over medium heat for 30 minutes.

* The shiitake mushrooms can be used right away after steamed. Or, you may keep them in freezer for later use.

三姐甘草欖角

Kitty's preserved black olives

材料

- 欖角 5 粒
- 果皮 1 小角
- 甘草 2 片
- 糖 1 茶匙
- 油 2 湯匙

做法

❶ 所有材料拌勻，大火蒸約 8-10 分鐘至軟身。使用前切粒。

* 可按比例增加各材料的份量製作大份量，待欖角攤涼後放入密封瓶內，放在陰涼處貯藏。緊記油一定要蓋過欖角。

Ingredients

- 5 preserved black olives
- 1 small piece dried tangerine peel
- 2 slices liquorice
- 1 tsp sugar
- 2 tbsp oil

Method

❶ Mix all ingredients in a steaming bowl. Steam over high heat until the olives are soft. Dice the mixture before using.

* You may size up the amounts listed proportionally to make a big batch. When the preserved black olives are cooled completely, transfer into an airtight container and keep in a cool spot away from the sun. Make sure there is enough oil to cover the olives.

基本烹調技巧

煲肉湯：肉類放入凍水內，待滾起後撈起、瀝乾，再用少量油在鑊中略煎，肉湯會更香濃。

煎魚：用大火燒熱鑊，下油，待油熱後，將魚放入鑊中，要煎至魚呈金黃色，才可以反轉再煎另一邊，待煎至金黃即可以上碟。

煎豆腐：如果想保持豆腐的完整，將豆腐浸入鹽水中，可以令豆腐結實，同時鹹味滲入豆腐中，味道會更佳。

炆肉：將肉類清洗後，用鹽醃一夜，抹去鹽粒後炆或煲湯皆可，令肉質更軟滑及好味。如配料中有八角、花椒、薑、豆瓣醬、柱侯醬等，宜先爆香才放入肉類，香料才能發揮效果滲入肉類中。

陳皮：柑皮儲藏三年以上方可叫陳皮。陳皮氣味芳香，蒸魚、肉餅或煲湯時下陳皮，有去腥增香的功效。

苦瓜

bitter melon

話梅梅酒涼拌白玉苦瓜

White bitter melon marinated in plum wine

材料

- 白玉苦瓜 1 條
- 話梅 15 粒
- 沸水 2 公升
- 梅酒 600 毫升
- 紫蘇葉 15 塊（新鮮及乾品）

Ingredients

- 1 white bitter melon
- 15 dried liquorice plums
- 2 litres boiling water
- 600 ml plum wine
- 15 dried and fresh Shiso leaves (a.k.a. perilla leaves)

品嘗三姐私房梅酒。
Trying the plum wine made by Chef Kitty Siu.

做法

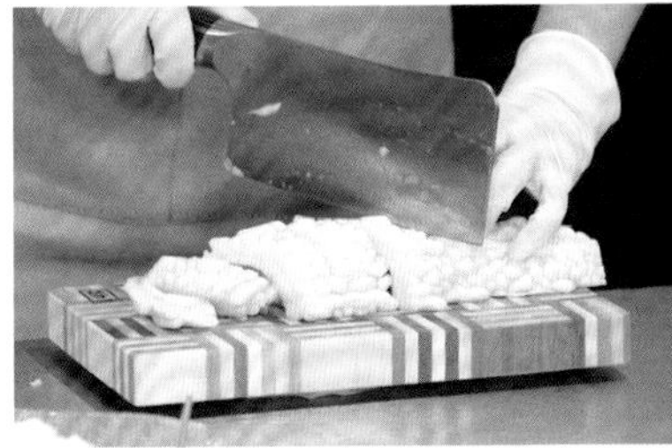

1. 預備一個 3 公升的玻璃瓶，放入話梅及乾紫蘇葉，倒入沸水（約 7 成滿），待完全涼透。
2. 苦瓜洗淨，切去兩端，橫切開半，刮掉瓜瓤，切條或片狀均可。
3. 玻璃瓶內添加梅酒，加入新鮮紫蘇葉（先用手揉搓），再放進苦瓜，蓋好並放雪櫃冷藏一晚，即可享用。

Method

1. Put dried plums and dried Shiso leaves into a 3-litre sealable jar. Pour in boiling water until the jar is 70% full. Leave it to cool completely.
2. Rinse the white bitter melon and cut off both ends. Cut in half across the length. Scoop out the seeds. Then cut into strips or slices.
3. Pour the plum wine into the glass jar. Rub to bruise the fresh Shiso leaves. Put the leaves into the glass jar. Then put in the white bitter melon. Seal the jar and refrigerate overnight. Serve.

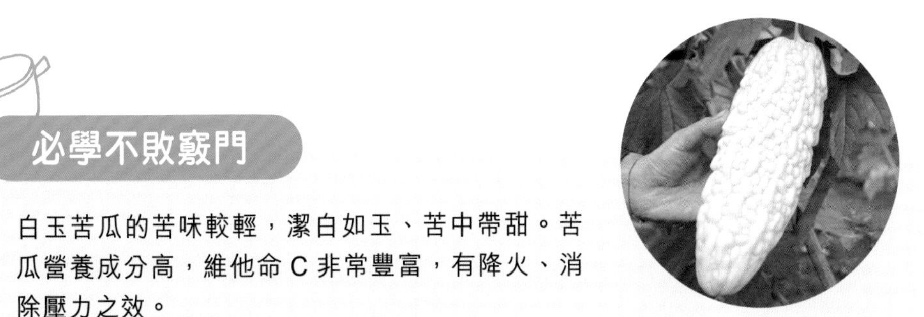

必學不敗竅門

- 白玉苦瓜的苦味較輕，潔白如玉、苦中帶甜。苦瓜營養成分高，維他命 C 非常豐富，有降火、消除壓力之效。
- White bitter melon is not as bitter as its green counterpart. It looks white and translucent with a hint of bitterness amid sweetness. Bitter melon is rich in nutrients, especially vitamin C. Chinese herbalists believe it clear the excessing heat and relieves stress.

OOKING TIPS

牛油果
avocado

芝士蟹肉牛油果杯

Baked avocado boat with cheese and crabmeat

必學不敗竅門

- 如使用焗爐，溫度及時間可稍自行調節，略為提升。
- 牛油果是超級食品；挑選時用手輕按，如感到微微軟熟有彈性，絕非死硬的，代表接近完熟可食用。
- If you're using a conventional oven, you have to bake them at a slightly higher temperature and maybe for slightly longer. You may have to experiment a bit.
- Avocado is a superfood. For the right ripeness, squeeze the avocado gently. If it yields to your finger and feels resilient, it is almost ripe and is ready to serve. If it is still rock hard, it's not ripe yet.

COOKING TIP

示範短片

材料

- 牛油果 1 個（大）
- 雞蛋 2 隻
- 即食蟹肉 2 湯匙
- 車厘茄 4 粒
- 甘筍薄片適量
- 芝士碎 2 湯匙
- 鹽適量
- 黑椒碎適量

Ingredients

- 1 large avocado
- 2 eggs
- 2 tbsp instant crabmeat
- 4 cherry tomatoes
- thinly sliced carrot
- 2 tbsp grated cheese
- salt
- ground black pepper

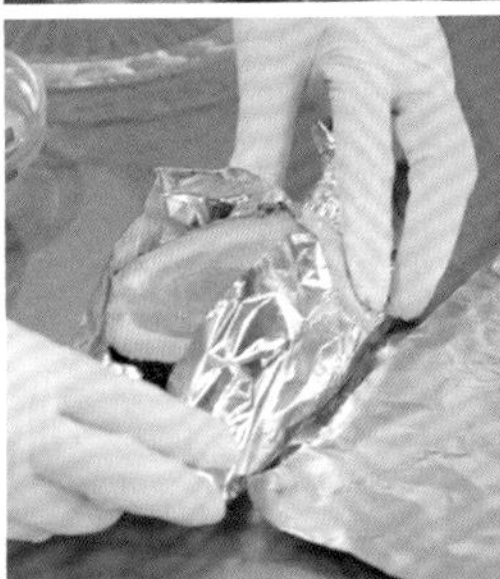

做法

1. 車厘茄一開八，備用。
2. 牛油果對切一半，去核，底部削平少許，在中間位置掏去少量果肉，再放入雞蛋。
3. 錫紙內放上甘筍薄片墊底防黏，放上牛油果後用錫紙包封。
4. 預熱氣炸鍋攝氏 160 度，放入牛油果焗 6 分鐘至蛋半熟，取出，放上車厘茄、即食蟹肉、芝士碎，灑上適量鹽、黑椒碎，再焗 3 分鐘即成。

Method

1. Cut each cherry tomato into eighths.
2. Cut avocado in half along the length. Remove the core. Cut a thin strip off the rounded bottoms of the avocado halves, so that they sit stably. Then scoop out some of the flesh and set aside. Crack an egg in each half.
3. Lay a piece of aluminium foil on the counter. Put a slice of carrot on the foil so that the avocado won't stick. Put the avocado over the carrot slice. Wrap it in foil. Repeat with the other avocado half.
4. Preheat an air-fryer to 160°C. Put in the avocado halves in aluminium foil. Cook for 6 minutes until the eggs are half-cooked. Remove avocado halves from the air-fryer. Add cherry tomatoes, instant crabmeat and grated cheese over the avocado boat. Sprinkle with salt and ground black pepper. Put them back into the air-fryer and cook for 3 minutes uncovered. Serve.

墨魚滑
minced cuttlefish

墨膠釀雞脯

Boneless chicken thigh stuffed with minced cuttlefish

必學不敗竅門

- 墨魚滑是用新鮮墨魚肉打製而成，口感彈牙，海水魚檔有售。
- 如雞扒太厚或厚薄不均必須先整理，再用刀劃數下。
- You can get minced cuttlefish from fishmongers selling marine fish. It has a springy texture after cooked.
- If the chicken thigh is too thick or uneven in thickness, please trim it to achieve even thickness first, before making a few light cuts on it.

COOKING TIP

材料

- 墨魚滑 200 克
- 雞扒 1 件
- 生粉 20 克
- 酸甜汁 3 湯匙（伴食）
- 日本紫蘇葉少量（裝飾用）
- 食用花少量（裝飾用）

Ingredients

- 200 g minced cuttlefish
- 1 boneless chicken thigh
- 20 g potato starch
- 3 tbsp sweet and sour sauce (as dressing)
- Japanese Shiso leaves (as garnish)
- edible flowers (as garnish)

做法

1. 雞扒洗淨，吸乾水分，用刀劃數下，以少許鹽略醃。
2. 墨魚滑在使用前先撻數下，均勻地鋪平在雞扒上，再蘸上生粉。
3. 熱鑊下油，放入釀雞扒（雞皮向下），慢火半煎炸至兩面金黃，盛起，瀝乾油分，切件。
4. 上碟後，可以食用花、紫蘇葉伴碟，淋上酸甜汁即成。

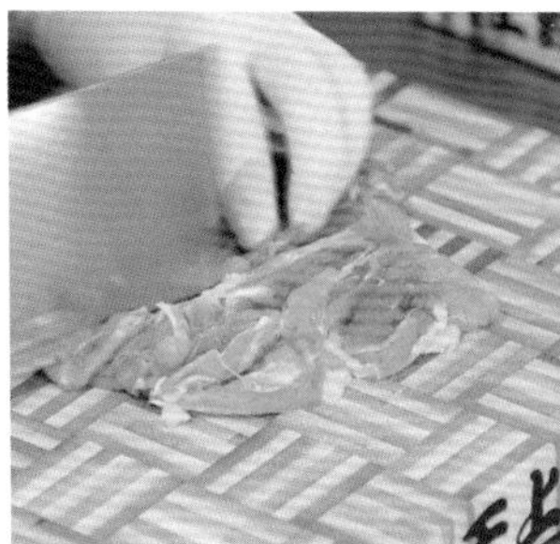

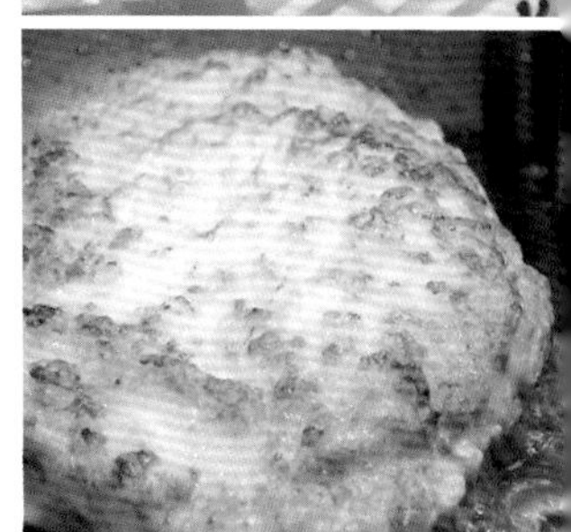

Method

1. Rinse the boneless chicken thigh and wipe dry. Make a few light cuts on the meat side. Add salt and mix well. Leave it to marinate briefly.
2. Before using store-bought minced cuttlefish, slap it forcefully onto a chopping board a few times to make it springy in texture. Then lay the chicken thigh flat on a counter with the meat side up. Smear the minced cuttlefish over the chicken meat evenly. Then coat both sides in potato starch.
3. Heat a wok and add oil. Put in the stuffed chicken thigh with the skin side down. Cook over low heat in semi-deep frying manner until both sides golden. Drain off excess oil. Cut into pieces.
4. Arrange on a serving plate. Garnish with edible flowers and Shiso leaves. Drizzle with sweet and sour sauce. Serve.

雞
chicken

電飯煲玫瑰露豉油雞

Soy-marinated chicken in electric rice cooker

材料

- 雞 1 隻
- 薑 4 片
- 葱絲少許

Ingredients

- 1 chicken
- 4 slices ginger
- finely shredded spring onion

調味料

- 玫瑰露 2 湯匙
- 甜豉油 8 湯匙
- 老抽 1 湯匙
- 糖 3 湯匙

Seasoning

- 2 tbsp Chinese rose wine
- 8 tbsp sweet soy sauce
- 1 tbsp dark soy sauce
- 3 tbsp sugar

做法

1. 雞洗淨，瀝乾水分，備用。
2. 將薑片及調味料放入電飯煲內，按煮飯模式，待醬汁滾起後放入雞，雞胸向下，加蓋。
3. 每 10-15 分鐘轉動雞一次，共 4 次，煮至雞全熟及雞皮上色，盛起放涼，斬件。
4. 將餘下醬汁煮至濃稠，淋上雞面，綴上葱絲即成。

Method

1. Rinse the chicken and drain well. Set aside.
2. Put ginger and seasoning into a rice cooker. Turn on the rice-cooking cycle. Cook until the mixture boils. Put the chicken in with the breast side down. Cover the lid.
3. Flip the chicken upside-down once every 10 to 15 minutes. After you flip the chicken for 4 times and when the chicken is cooked through and evenly coloured, remove from the rice cooker to let cool. Chop into pieces and arrange on a serving plate.
4. Cook to reduce the remaining sauce. Drizzle over the chicken. Garnish with finely shredded spring onion. Serve.

必學不敗竅門

- 待醬汁煮滾才放入雞，雞肉不易霉。
- 因雞胸肉厚，排入電飯煲時建議雞胸向下，容易熟透。
- Heat the sauce till it boils before putting the chicken in. The chicken flesh is less likely to turn mushy this way.
- The breast is the fleshiest part of a chicken. It is advisable to put the chicken into the rice cooker with the breast side down, so that it can be cooked through more easily.

雞
chicken

海南雞飯
Hainan chicken rice

做法

1. 雞洗淨、抹乾水分，以粗鹽塗抹雞內外，淋上紹興酒，放雪櫃醃一晚。
2. 香茅洗淨，去外皮，取芯切段，拍扁；蒜肉去皮，原粒備用。
3. 雞醃好後，隔水蒸 25-30 分鐘，蒸好後浸泡冰水降溫，蒸雞汁留用，雞冷卻後斬件上碟。
4. 熱鑊下油，爆香薑片，再放入蒜肉、香茅、已洗淨米粒爆炒至香味溢出，倒進電飯煲，加入雞湯及蒸雞汁，加蓋按掣煮飯。
5. 預備薑蓉：葱蓉、薑蓉及生抽混和，淋上熟油即成。
6. 飯煮熟後，取走薑片、蒜肉及香茅，盛起飯，排好雞件，伴以薑蓉享用。

必學不敗竅門

- 雞蒸熟後即泡冰水，令雞皮收縮，肉質爽滑。
- After the chicken is steamed, dunk it into a bowl of ice water. The thermal shock would make the skin springy and the flesh velvety.

COOKING TIP

材料

- 雞 1 隻
- 雞湯 3 量米杯
- 米 3 量米杯
- 薑片 15-20 片（拍扁）
- 蒜肉 8-10 瓣
- 香茅 2-3 條
- 粗鹽 1-2 湯匙
- 紹興酒 2 湯匙

Ingredients

- 1 chicken
- 3 cups* chicken stock
- 3 cups* rice
- 15 to 20 slices ginger (crushed)
- 8 to 10 cloves garlic
- 2 to 3 sprigs lemongrass
- 1 to 2 tbsp coarse salt
- 2 tbsp Shaoxing wine

示範短片

薑蓉蘸料

- 薑蓉 2 湯匙
- 葱蓉 5 湯匙
- 生抽 1 湯匙
- 熟油 2-3 湯匙

Dipping sauce

- 2 tbsp grated ginger
- 5 tbsp finely chopped spring onion
- 1 tbsp light soy sauce
- 2 to 3 tbsp cooked oil

* use the cup that comes with the rice cooker to measure

Method

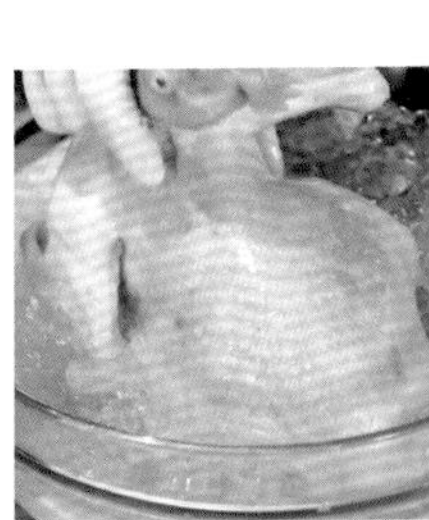

1. Rinse the chicken and wipe dry. Rub coarse salt on both the inside and the outside of the chicken evenly. Drizzle with Shaoxing wine and rub again. Leave it to marinate in the fridge overnight.
2. Rinse the lemongrass and peel off the dry leaves outside. Cut into short lengths and bruise them with the back of a knife. Set aside. Peel the garlic cloves.
3. Take the chicken out of the fridge and put it in a steaming dish. Steam for 25 to 30 minutes. Then dunk the chicken into ice water to cool it off rapidly. Reserve the juices in the steaming dish for later use. When the chicken is cool to the touch, chop it up into pieces and arrange on a serving plate.
4. To make the rice, heat a wok and add oil. Stir-fry ginger until fragrant. Put in garlic cloves, lemongrass and rice that has been rinsed. Toss the mixture until fragrant. Transfer the mixture into a rice cooker. Add chicken stock and the juices from steaming chicken (from step 3). Turn on the rice-cooking programme.
5. To make the dip, mix ginger, spring onion and soy sauce together. Heat some oil until smoking hot. Drizzle over the ginger, spring onion and soy sauce mixture.
6. After the rice cooker completes the programme, remove the ginger, garlic and lemongrass. Fluff the rice and scoop it into serving bowls. Serve the chicken with the rice and the dip on the side.

茄子
eggplant

銀魚乾蒸茄子

Steamed eggplant with dried anchovies

材料

- 幼身茄子 3 條
- 銀魚乾 40 克
- 豬腩肉 160 克
- 蒜粒 1 湯匙
- 指天椒 1 隻（切圈）
- 葱花少許

Ingredients

- 3 Japanese eggplants
- 40 g dried anchovies
- 160 g pork belly
- 1 tbsp diced garlic
- 1 bird's eye chilli (cut into rings)
- finely chopped spring onion

醬料

- 甜麵醬 1 茶匙
- 蠔油 1/2 茶匙

Sauce

- 1 tsp sweet bean paste
- 1/2 tsp oyster sauce

做法

1. 茄子洗淨，原條橫切半，放入水內，加白醋 1 湯匙浸泡備用。
2. 茄子瀝乾水分，隔水蒸 7-10 分鐘，備用。
3. 銀魚乾洗淨、浸軟，抹乾水分。熱鑊下油，爆香銀魚乾至金黃，盛起備用。
4. 腩肉洗淨，切小粒，加適量油、鹽、生抽、麻油醃勻。熱鑊下油，爆香肉粒，下蒜粒、指天椒爆炒，加入甜麵醬及蠔油炒勻，盛起鋪上茄子上，再放上銀魚乾，加蓋再蒸 2-3 分鐘，最後撒上葱花即成。

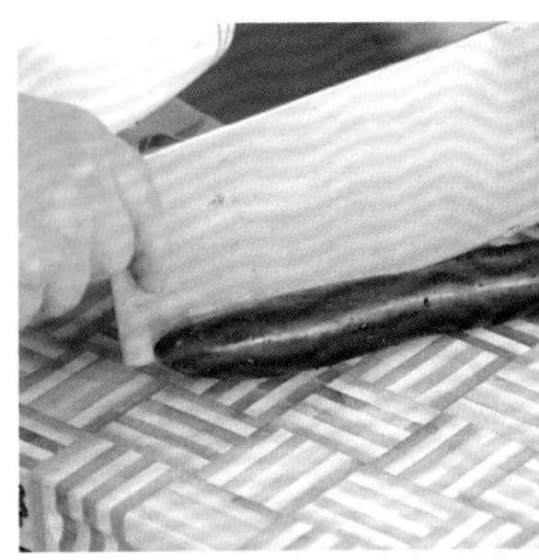

Method

1. Rinse the eggplants. Cut each in half across the length. Put them in a bowl of water and add 1 tbsp of white vinegar.
2. Drain the eggplants. Steam for 7 to 10 minutes. Set aside.
3. Rinse the dried anchovies. Soak them in water till soft. Drain and wipe dry. Heat a wok and add oil. Stir-fry the dried anchovies until golden. Drain and set aside.
4. Rinse the pork belly. Finely dice it and put into a bowl. Add oil, salt, light soy sauce and sesame oil. Mix well and leave it briefly. Heat a wok and add oil. Stir-fry the diced pork until fragrant. Add diced garlic and bird's eye chilli. Toss well. Add sweet bean paste and oyster sauce. Toss again. Pour the mixture over the steamed eggplants from step 2. Arrange dried anchovies on top. Steam for 2 to 3 minutes. Sprinkle with spring onion. Serve.

必學不敗竅門

- 浸泡茄子時加入少許白醋，令茄子肉又白又滑。
- Soaking the eggplants in water with a dash of white vinegar prevents the eggplant from discolouring and keeps it velvety in texture.

涼瓜
bitter melon

土魷絲炆釀涼瓜

Braised stuffed bitter melon with dried squid

必學不敗竅門

- 釀涼瓜時，餡料必須蘸上生粉才緊黏瓜肉，不容易散開。
- Before you stuff the bitter melon with the filling, you must coat the filling in potato starch which acts as a binding agent. Otherwise, the filling may fall out of the bitter melon in the cooking process.

COOKING TIP

材料

- 沖繩苦瓜 1 條
- 鯪魚肉 200 克
- 免治豬肉 100 克
- 乾土魷 2 隻
- 莧菜 100 克
- 馬友鹹魚肉 1 湯匙
- 蒜粒 2 湯匙
- 紅尖椒絲少量
- 胡椒粉少許
- 蝦米粉 1 湯匙

Ingredients

- 1 Okinawa bitter melon
- 200 g minced dace
- 100 g ground pork
- 2 dried squids
- 100 g amaranth
- 1 tbsp salted fish (threadfin)
- 2 tbsp diced garlic
- finely shredded red chillies
- ground white pepper
- 1 tbsp ground dried shrimp

芡汁

- 蠔油 1 茶匙
- 魚露 1/2 茶匙
- 生抽 1/2 茶匙
- 生粉水適量

Glaze

- 1 tsp oyster sauce
- 1/2 tsp fish sauce
- 1/2 tsp light soy sauce
- potato starch thickening glaze

沖繩苦瓜味道甘苦；外形比中國苦瓜有更多凹凸表層。

Okinawa bitter melon is bitter in taste. It has more bumps on the skin than its Chinese counterparts.

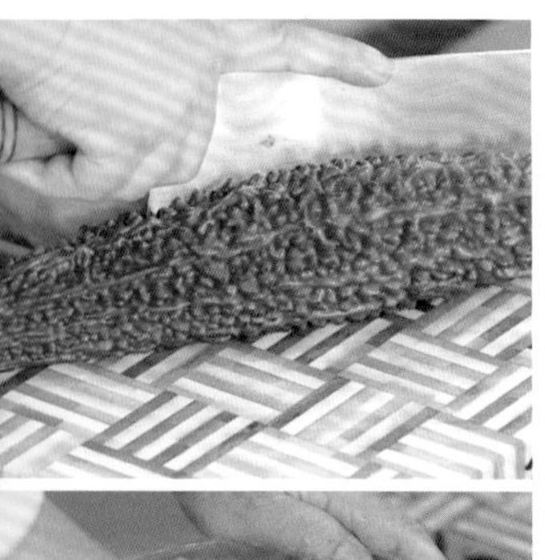

做法

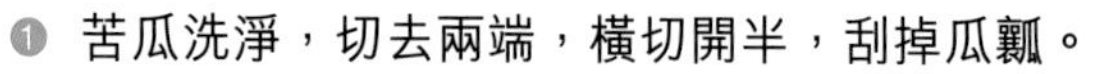

❶ 苦瓜洗淨，切去兩端，横切開半，刮掉瓜瓤。
❷ 莧菜洗淨、去根，用熱水汆燙，瀝乾後切粒備用。
❸ 鹹魚肉切小粒；乾土魷洗淨、浸軟，去皮、切絲備用。
❹ 鯪魚肉加入免治豬肉、鹹魚粒、莧菜粒，下少許胡椒粉、蝦米粉 1 湯匙調味，順方向把所有材料撻勻，並撻至起膠，取適量蘸上生粉，直接釀入苦瓜內，均勻釀滿後在表面撲上適量生粉。
❺ 熱鑊下油，下苦瓜（肉向底）煎至金黃。
❻ 土魷絲放入鑊內煎香，下蒜粒及水 100 毫升炆約 5 分鐘。
❼ 苦瓜盛起，加入芡汁煮勻，下紅椒絲，淋於苦瓜上即成。

Method

❶ Rinse the bitter melon. Cut off both ends. Cut in half across the length. Scoop out the seeds.
❷ Rinse the amaranth and cut off the roots. Blanch in boiling water. Drain and dice.
❸ Finely dice the salted fish. Set aside. Rinse the dried squids and then soak them in water till soft. Remove the purple skin and finely shred them.
❹ Put minced dace into a mixing bowl. Add ground pork, diced salted fish, amaranth, ground white pepper and 1 tbsp of ground dried shrimp. Stir in one direction to mix well. Lift the mixture off the bowl and slap it back in forcefully. Repeat lifting and slapping until sticky. Scoop out some of the mixture and coat it in potato starch. Stuff it into the bitter melon until full. Coat the surface of the filling with potato starch once more.
❺ Heat wok and add oil. Put in the stuffed bitter melon with the filling facing down. Fry until golden.
❻ Add the dried squids. Fry until fragrant. Add diced garlic and 100 ml of water. Bring to the boil and cover the lid. Cook for 5 minutes.
❼ Remove the bitter melon and arrange on a serving plate. Add the glaze ingredients to the wok and stir well. Add red chillies and toss again. Drizzle the glaze over the bitter melon. Serve.

牛柳

beef tenderloin

開胃椒醬豬牛雙併

Stir-fried beef tenderloin and pork with fermented soy paste

必學不敗竅門

- 醃牛柳時，先加入糖、生粉及油拌勻，以免生抽鹹味太快滲入牛柳內而過鹹。
- 炸花生不要與其他材料拌炒，否則花生會太腍而不香脆。
- When you marinate the beef, add sugar, potato starch and oil first. Stir well. If you put in the light soy sauce first, the beef will pick it up too quickly and may become too salty.
- Do not stir-fry the peanuts with other ingredients. Otherwise, the peanuts won't be crispy.

COOKING TIPS

材料

- 牛柳 300 克
- 蝦米 12 隻
- 冬菇 3 朵
- 菜脯 2 條（切粒）
- 青、紅甜椒各 1/3 個
- 西芹 100 克（切粒）
- 五香豆乾 2 件（切粒）
- 乾辣椒 3-4 隻
- 肥豬肉 200 克（炸豬油渣用）
- 蒜粒 1 湯匙
- 炸花生少許

調味料

- 磨豉醬 1 茶匙
- 麵豉醬 1 茶匙
- 生抽 1 茶匙
- 紹興酒 1 湯匙
- 糖 1/4 茶匙

Ingredients

- 300 g beef tenderloin
- 12 dried shrimps
- 3 dried shiitake mushrooms
- 2 strips dried radish (diced)
- 1/3 green bell pepper
- 1/3 red bell pepper
- 100 g celery (diced)
- 2 pieces five-spice dried tofu (diced)
- 3-4 dried red chillies
- 200 g fatty pork (to be fried into cracklings)
- 1 tbsp diced garlic
- deep-fried peanuts

Seasoning

- 1 tsp ground bean paste
- 1 tsp fermented soy paste
- 1 tsp light soy sauce
- 1 tbsp Shaoxing wine
- 1/4 tsp sugar

做法

1. 牛柳切粒，加入糖約 1/2 茶匙、生粉 1 茶匙及油 1 湯匙拌勻，再下生抽 1 茶匙拌勻略醃，備用。
2. 蝦米浸軟，瀝乾水分（水留用），下油鑊爆香，盛起備用。
3. 冬菇浸軟，瀝乾水分，去蒂，切粒；青紅椒去籽，切粒備用。
4. 製作豬油渣：肥豬肉洗淨，飛水，切粒。放入鑊內，加入少許水，以中火慢慢煎香，期間不停攪拌，待豬油逐漸釋出，豬肉粒收至乾脆成豬油渣，盛起備用（豬油可留用）。
5. 熱鑊下油，煎香牛柳粒至金黃半熟，盛起。
6. 熱鑊下油，爆香蒜粒，下五香豆乾粒、冬菇粒、菜脯粒、西芹粒及青紅椒粒炒勻，下適量蝦米水，加入麵豉醬、磨豉醬爆香，炒勻後下蝦米、乾辣椒、豬油渣，加入餘下的調味料拌勻，最後牛柳粒回鑊炒勻，上碟放上炸花生即成。

Method

1. Dice the beef tenderloin. Add 1/2 tsp of sugar, 1 tsp of potato starch and 1 tbsp of oil. Mix well. Add 1 tsp of light soy sauce. Mix again and leave it briefly.
2. Soak the dried shrimps in water till soft. Drain and set aside the soaking water. Stir-fry dried shrimps in a wok with some oil until fragrant. Set aside.
3. Soak shiitake mushrooms in water till soft. Drain well. Cut off the stems and dice them. Set aside. De-seed the red and green bell peppers. Dice them.
4. To make cracklings, rinse the pork and blanch it in boiling water. Drain. Dice it and put into a wok. Add a little water and cook over medium heat until browned. Keep stirring throughout the cooking process. When all fat is rendered and the diced pork turn into crispy dry bits, drain the lard and set aside the cracklings. (You may save the lard for making other dishes.)
5. Heat wok and add oil. Fry the beef tenderloin until golden and half-cooked. Set aside.
6. Heat wok again and add oil. Stir-fry garlic until fragrant. Add five-spice dried tofu, shiitake mushrooms, dried radish, celery, red and green peppers. Toss well. Pour in the soaking water of dried shrimps. Add ground bean paste and fermented soy paste. Toss well. Add dried shrimps, dried red chillies and pork cracklings from step 4. Add the remaining seasoning. Put the beef tenderloin back in the wok. Toss well and save on a serving plate. Sprinkle with deep-fried peanuts on top. Serve.

豬板筋

pork silver skin

甜梅菜青豆炒豬板筋

Stir-fried pork silver skin with sweet Mei Cai and green string beans

材料

- 豬板筋 250 克
- 甜梅菜 1/2 棵
- 青豆角 10 條
- 蒜蓉 2 湯匙
- 蠔油 1 茶匙
- 紹興酒 1 湯匙
- XO 醬 1 茶匙
- 油葱酥 1 湯匙

Ingredients

- 250 g pork silver skin
- 1/2 sprig sweet Mei Cai (salted mustard greens)
- 10 pods green string beans
- 2 tbsp grated garlic
- 1 tsp oyster sauce
- 1 tbsp Shaoxing wine
- 1 tsp XO sauce
- 1 tbsp deep-fried shallot bits

做法

❶ 豬板筋洗淨，切片，加少許生粉、糖、生抽、麻油醃勻。
❷ 青豆角洗淨，切段。
❸ 甜梅菜洗淨，浸水，榨乾水分及吸乾，切粒，以白鑊炒香備用。
❹ 熱鑊下油，以中小火爆香蒜蓉，下豬板筋爆炒至 8 成熟，加入青豆角、XO 醬及甜梅菜炒熟；下蠔油，灒酒炒勻，上碟，撒上油葱酥即成。

Method

❶ Rinse the pork and slice it. Add a pinch of potato starch, sugar, light soy sauce and sesame oil. Mix well.
❷ Rinse the green string beans. Cut into short lengths.
❸ Rinse the sweet Mei Cai. Soak it in water for a while. Drain, squeeze dry and wipe dry. Dice it and stir-fry in a dry wok until fragrant.
❹ Heat wok and add oil. Stir-fry garlic over medium-low heat until fragrant. Put in the pork and stir-fry until medium-well done. Add green string beans, XO sauce and sweet Mei Cai. Toss until the pork is cooked through. Add oyster sauce and Shaoxing wine. Toss again. Transfer onto a serving plate. Sprinkle with deep-fried shallot bits on top. Serve.

必學不敗竅門

- 梅菜浸泡及瀝乾水分後，用白鑊烘香至微焦，會令味道更突出。
- 豬板筋是里脊肉相連的一層筋膜，炒熟後非常嫩滑。
- After soaking and draining the Mei Cai, stir-frying it in a dry wok until lightly browned to intensify its flavours.
- Pork silver skin is a membrane of connective tissue on a pork loin. It is soft but with a lovely chew after cooked.

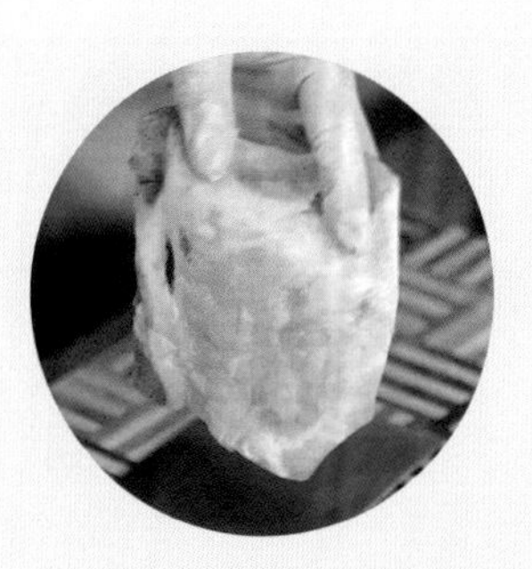

COOKING TIPS

豬

pork

沙葛蒜心炒鹹豬肉

Stir-fried salted pork belly with yam bean and garlic sprouts

材料

- 五花腩 450 克
- 沙葛 1/2 個
- 蒜心 8 條
- 土魷 1 隻
- 鮑貝 10 粒
- 鮮百合 1 球
- 蒜粒 1 湯匙
- XO 醬 2 茶匙
- 紅椒絲少許

Ingredients

- 450 g pork belly
- 1/2 yam bean
- 8 sprigs garlic sprouts
- 1 dried squid
- 10 Pacific clams
- 1 fresh lily bulb
- 1 tbsp diced garlic
- 2 tsp XO sauce
- finely shredded red chillies

做法

❶ 五花腩洗淨，以粗鹽 1 湯匙醃一晚成鹹肉。使用前抹去鹽粒，隔水蒸 7-8 分鐘，取出後浸冰水，切片備用。

❷ 沙葛洗淨，去皮，切條；鮮百合洗淨，拆成瓣；蒜心洗淨，切段。

❸ 土魷洗淨，浸軟，斜刀劃紋，翻轉後切片，觸鬚切半。

❹ 熱鑊下油，大火煎香鹹肉至兩面金黃，下蒜粒略爆香，下土魷、鮑貝、沙葛、蒜心、百合、XO 醬爆炒，灒入適量紹興酒，最後下少許糖、紅椒絲炒勻，上碟即成。

Method

❶ Rinse the pork. Rub 1 tbsp of coarse salt over it evenly. Leave it overnight. Remove all the salt on the surface the next day. Steam for 7 to 8 minutes. Soak in ice water until cool. Slice and set aside.

❷ Rinse the yam bean. Peel and cut into strips. Rinse the fresh lily bulb. Break into scales and set aside. Rinse the garlic sprouts. Cut into short lengths.

❸ Rinse the dried squid. Soak in water till soft. Make light criss-cross cut on one side. Flip it over and slice it. Cut the tentacles in half.

❹ Heat wok and add oil. Stir-fry salted pork from step 1 over high heat until both sides golden. Add garlic and toss until fragrant. Put in dried squid, Pacific clams, yam bean, garlic sprouts, lily bulb and XO sauce. Toss to mix well. Drizzle with Shaoxing wine. Add sugar and red chillies at last. Toss again and save on a serving plate. Serve.

必學不敗竅門

- 五花腩蒸後，即放入冰水或雪櫃冷藏片刻，切片效果更平均、更完整。
- After steaming the pork belly, soak it in ice water or refrigerate for a while. That would make the pork firmer in texture and can be sliced more easily and evenly.

COOKING TIPS

豬肋條
pork rib

甜薑京蔥炆豬肋條

Braised pork ribs with candied ginger and scallion

材料	Ingredients
西班牙豬肋條 500 克	500 g Iberico pork ribs (cut into strips)
老薑 80 克	80 g old ginger
子薑 80 克	80 g young ginger
紅糖 4 湯匙	4 tbsp light brown sugar
甜麵醬 1 茶匙	1 tsp sweet soybean paste
柱侯醬 2 茶匙	2 tsp Chu Hau sauce
京蔥 2 條	2 Peking scallions
紹興酒 100 毫升	100 ml Shaoxing wine

做法

1. 子薑及老薑洗淨，瀝乾水分，連皮切厚片，用紅糖分別醃過夜；老薑用前先拍扁。
2. 京葱一半斜切粗片，一半切絲，備用。
3. 豬肋條洗淨，切段，加入蛋白半隻、鹽少許及麵粉 1 湯匙略醃片刻，備用。
4. 熱鑊下油，爆香京葱絲，盛起上碟圍邊；原鑊爆香京葱片至微焦，盛起。
5. 燒熱煲仔，下油燒熱後加入甜老薑片爆香，下豬肋條煎至兩面金黃色，再下甜子薑片，加甜麵醬及柱侯醬爆香，煎煮至乾身，灒入紹興酒，加清水 100 毫升，炆約 10 分鐘後，下京葱片炒至收汁（如嗜辣可加入辣椒），上碟即成。

Method

1. Rinse both old and young ginger. Drain well. Slightly crush the old ginger with the flat side of a knife. Slice both thickly with skin on, but keep them separate. Add light brown sugar and mix well. Leave them overnight.
2. Divide the Peking scallions into half. Slice half thickly and diagonally. Finely shred the rest. Set aside separately.
3. Rinse the pork ribs. Cut into short lengths. Add 1/2 an egg white, a pinch of salt and 1 tbsp of flour. Mix well and leave them briefly.
4. Heat wok and add oil. Stir-fry finely shredded Peking scallions until fragrant. Transfer onto a serving plate and arrange along the rim of the plate. In the same wok, stir-fry the thickly sliced Peking scallions until lightly browned. Set aside.
5. Heat a clay pot. Add oil and heat it up. Put in the candied old ginger. Fry until fragrant. Put in the ribs and fry until both sides golden. Add candied young ginger, sweet soybean paste and Chu Hau sauce. Toss well and stir until the juices reduce. Drizzle with Shaoxing wine. Add 100 ml of water. Cover the lid and cook for 10 minutes. Put in the thickly sliced Peking scallions, and optionally sliced red chilli. Toss until the sauce reduces. Serve.

必學不敗竅門

- 每片薑都要爆香後才放入豬肋條。
- 炆豬肋條前加適量蛋白及麵粉，會令豬肋條更易掛汁，亦有收汁的效果。
- Make sure you fry every slice of candied old ginger until lightly browned before putting in the ribs.
- Coating the pork ribs in egg white and flour helps the sauce cling on them. It will also take less time to reduce the sauce.

大白菜
bok choy

蝦乾拍薑煮大白菜

Braised Bok Choy with dried shrimps and ginger

材料

大白菜 500 克
蝦乾 40 克
雞蛋 3 隻
炸枝竹 2 條
薑 8 片
上湯 200 毫升
胡椒粉 1/2 茶匙

Ingredients

- 500 g Bok Choy
- 40 g large dried shrimps
- 3 eggs
- 2 deep-fried beancurd sticks
- 8 slices ginger
- 200 ml stock
- 1/2 tsp ground white pepper

做法

❶ 蝦乾沖洗後浸軟，吸乾備用。
❷ 薑片拍鬆；炸枝竹用暖水浸軟，汆水備用。
❸ 薑片放入油鑊爆香，待薑片金黃後，加入雞蛋（免拂）煎至兩面熟透，盛起備用。
❹ 原鑊加入蝦乾，轉小火烘香，加入大白菜、上湯及水適量，轉中大火，加蓋煮 3 分鐘。
❺ 開蓋，稍為翻炒，加入煎蛋（分開成數份）及炸枝竹，加入胡椒粉煮開，上碟即成。

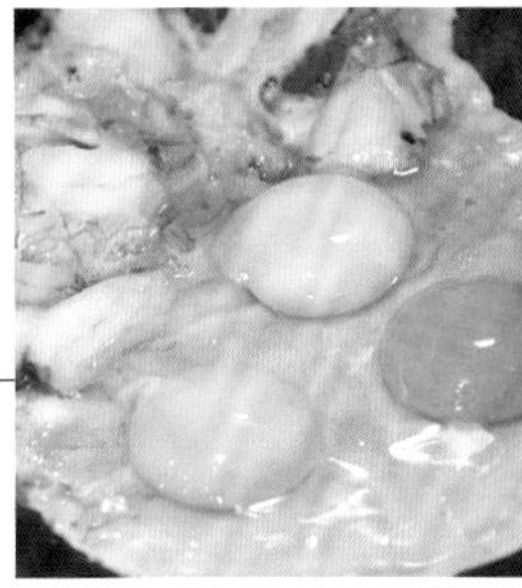

Method

❶ Rinse the dried shrimps and soak them in water till soft. Drain and wipe dry.
❷ Crush the sliced ginger with the flat side of a knife. Set aside. Soak deep-fried beancurd sticks in warm water till soft. Scald and set aside.
❸ Heat wok and add oil. Stir-fry ginger until fragrant and lightly browned. Crack the eggs in and do not stir or whisk the eggs. Fry till both sides cooked through. Set aside. Cut into a few pieces.
❹ In the same wok, stir-fry the dried shrimps over low heat until fragrant. Add Bok Choy, stock and some water. Turn to medium-high heat. Cover the lid and cook for 3 minutes.
❺ Open the lid and toss Bok Choy briefly. Put in fried eggs and deep-fried beancurd sticks. Sprinkle with ground white pepper. Bring to the boil again. Serve.

必學不敗竅門

- 大白菜較寒涼，可加多些薑同煮。
- 蝦乾用小火烘香，可去除腥味。
- Bok Choy is said to be Cold in nature from Chinese medical point of view. It's customary to add more ginger to balance out the Coldness.
- Frying dried shrimps over low heat helps remove the fishy taste.

OOKING TIPS

花膠
fish maw

花膠肉絲燴芽菜

Braised mung bean sprouts with fish maw and shredded pork

材料

花膠 400 克（已浸發）
芽菜 200 克
韭黃 120 克（切段）
韭菜花 120 克（切段）
脢頭肉 200 克
蒜粒 1 湯匙
甘筍絲少許
紅椒絲少許
紹興酒適量
蠔油 1 茶匙

Ingredients

- 400 g fish maw (rehydrated)
- 200 g mung bean sprouts
- 120 g yellow chives (cut into short lengths)
- 120 g flowering chives (cut into short lengths)
- 200 g pork shoulder butt
- 1 tbsp diced garlic
- shredded carrot
- shredded red chillies
- Shaoxing wine
- 1 tsp oyster sauce

做法

❶ 花膠切絲，用淡鹽水浸泡備用。
❷ 脢頭肉洗淨、切絲，加入油 1 茶匙、糖少許、生粉 1 茶匙、生抽 2 茶匙拌勻，略醃備用。
❸ 燒熱油鑊，爆香部分蒜粒，放入芽菜大火快炒，灒入紹興酒，加入韭黃及韭菜花快炒至熟，再放入甘筍絲及紅椒絲，放入鹽適量、糖少許炒勻後盛起，瀝去水分。
❹ 燒熱油鑊，放入肉絲爆炒，加入花膠絲及蒜粒 1 茶匙爆炒至肉絲熟透，灒入紹興酒，以少許鹽調味，將③回鑊，加入蠔油炒勻，上碟即成。

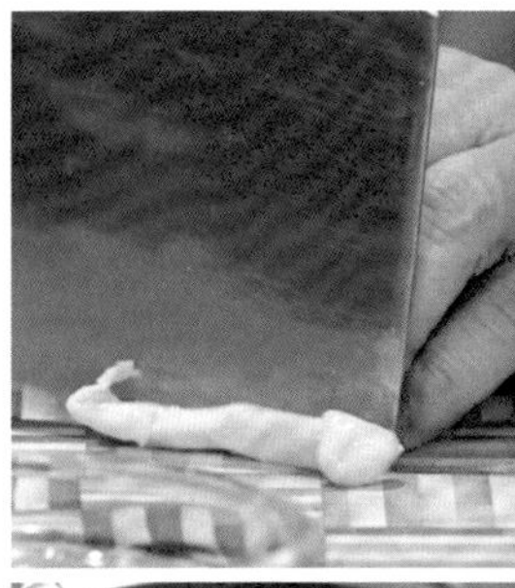

Method

❶ Shred the fish maw. Soak it in lightly salted water. Set aside.
❷ Rinse the pork and shred it. Add 1 tsp of oil, a pinch of sugar, 1 tsp of potato starch and 2 tsp of light soy sauce. Mix well and leave it briefly.
❸ Heat wok and add oil. Stir-fry some garlic until fragrant. Put in mung bean sprouts and toss quickly over high heat. Drizzle with Shaoxing wine. Add yellow chives and flowering chives. Toss quickly until cooked. Add carrot, red chillies, a pinch of salt and sugar. Toss to mix well. Set the mixture aside. Drain off any liquid.
❹ Heat wok again and add oil. Stir-fry the pork over high heat. Drain the fish maw and add to the wok. Add 1 tsp of diced garlic. Toss until pork is cooked through. Drizzle with Shaoxing wine and season with a pinch of salt. Put the mung bean sprouts mixture from step 3 back in the wok. Add oyster sauce. Toss well. Serve.

必學不敗竅門

- 芽菜要炒得好，炒後仍有爽脆口感，在炒前須將芽菜印乾水分，並用熱鑊滾油兜炒。
- For the best result, the mung bean sprouts should still retain a lovely crunch after stir-fried. Make sure you wipe the mung bean sprouts dry before stir-frying, and make sure the wok and oil are scalding hot.

夏枯草
xia ku cao

夏枯草淡菜煲豬腱

Pork shin soup with Xia Ku Cao and dried mussels

材料

- 夏枯草 120 克
- 淡菜 120 克
- 石決明 80 克
- 豬腱（老鼠腱）450 克
- 蜜棗 4 粒
- 果皮 1 瓣

Ingredients

- 120 g Xia Ku Cao
- 120 g dried mussels
- 80 g Shi Jue Ming (abalone shell)
- 450 g pork shin
- 4 candied dates
- 1 slice dried tangerine peel

做法

1. 夏枯草洗淨；淡菜浸洗數遍至完全乾淨；果皮浸軟，備用。
2. 石決明沖洗乾淨，放入茶袋備用。
3. 燒開水，放入石決明及蜜棗煮滾。
4. 豬腱汆水，撈起，切數刀以便更易釋出味道。
5. 燒熱油鑊，放入豬腱及淡菜煎香兩面，加入少許熱水沖走油分，夾起豬腱及淡菜放入湯鍋內，煲半小時。
6. 加入夏枯草及果皮，繼續煲 20 分鐘，完成。

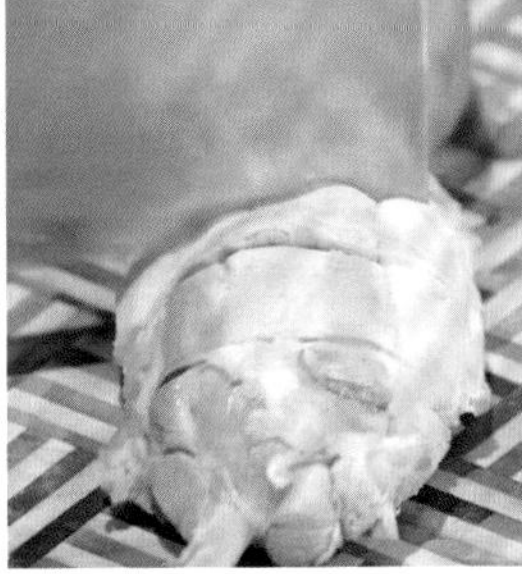

Method

1. Rinse the Xia Ku Cao. Drain well and set aside. Rinse the dried mussels a few times until the water runs clear. Drain and set aside. Soak the dried tangerine peel in water till soft. Set aside.
2. Rinse Shi Jue Ming well. Put into a tea bag. Set aside.
3. Boil water in a soup pot. Put in Shi Jue Ming and candied dates. Bring to the boil.
4. Blanch pork shin in boiling water. Drain and make a few cuts on the pork to let the flavours release more readily.
5. Heat wok and add oil. Put in pork shin and dried mussels. Fry until both sides of the pork turn golden. Pour in some hot water to rinse off the grease. Transfer the pork and dried mussels into the soup pot. Boil for 30 minutes.
6. Add Xia Ku Cao and dried tangerine peel. Cook for 20 minutes. Serve.

必學不敗竅門

- 石決明是鮑魚的殼，在藥材舖有售。為免飲湯時有碎屑，煲湯時一定要將石決明放入茶袋內，待湯完成後將茶袋棄去。
- 豬腱汆水後於表面切幾刀再略煎片刻，肉味會更香濃。
- Shi Jue Ming is abalone shell and you can get it from Chinese herbal store. To avoid having chipped shells in the soup that may cut your mouth, you must put Shi Jue Ming into a tea bag or muslin bag before using. Remove the tea bag or muslin bag when the soup is done.
- After blanching the pork shin, make a few cuts on it and fry it briefly in oil. That would make the meaty flavour stronger in the soup.

牛柳
beef tenderloin

薑汁酒芥蘭牛

Stir-fried sliced beef with kale in ginger wine

材料

- 芥蘭苗 400 克
- 牛柳 250 克
- 薑 50 克（榨汁用）
- 薑 6 片（拍扁）
- 玫瑰露 1 茶匙
- 紹興酒適量
- 紅椒少許（切角）
- 芹菜少許（切段，可省卻）

Ingredients

- 400 g baby Chinese kale
- 250 g beef tenderloin
- 50 g ginger (to be squeezed into juice)
- 6 slices ginger (crushed)
- 1 tsp Chinese rose wine
- Shaoxing wine
- red chillies (cut into wedges)
- Chinese celery (cut into short lengths, optional)

做法

❶ 薑磨蓉後榨汁，約 2 湯匙即可，備用。
❷ 牛柳去筋，逆紋切薄片，放入糖 1/4 茶匙、生粉 1 茶匙、油 1 茶匙及生抽 1/2 湯匙醃勻，備用。
❸ 燒熱油鑊，先爆香半份薑片，放入芥蘭苗大火快炒，待芥蘭苗炒香變軟，加入片糖碎 1 茶匙（砂糖亦可）、鹽少許炒勻，灒入薑汁、玫瑰露及紹興酒炒勻，盛起，瀝乾備用。
❹ 燒熱油鑊，爆香餘下薑片，用筷子將牛柳撥散，加入鑊內煎香至微焦即灒入紹興酒，放入芥蘭苗快速炒勻，下紅椒角及芹菜段（可省卻）兜勻，上碟即成。

Method

❶ Grate the 50 g of ginger and squeeze the juice out. You'd need about 2 tbsp for this recipe.
❷ Trim off the sinews on the beef. Slice thinly across the grain. Add 1/4 tsp of sugar, 1 tsp of potato starch, 1 tsp of oil and 1/2 tbsp of light soy sauce. Mix well.
❸ Heat wok and add oil. Stir-fry 3 slices of ginger until fragrant. Add kale and toss quickly over high heat till the kale wilt and softens. Add 1 tsp of crushed raw cane sugar slab (or sugar) and a pinch of salt. Toss well. Drizzle with ginger juice, Chinese rose wine and Shaoxing wine. Toss again. Set aside. Drain off the liquid.
❹ Heat wok and add oil. Stir-fry the remaining sliced ginger until fragrant. Separate the beef slices with chopsticks and put them into the wok. Fry until lightly browned. Drizzle with Shaoxing wine and put the kale back in. Toss quickly. Add red chillies and Chinese celery (optional). Toss and serve.

必學不敗竅門

- 牛柳買回來先放雪櫃冷藏一會，較容易切成片，而且肉質亦會鬆軟一點。
- 炒芥蘭苗時加入片糖碎，除令芥蘭更翠綠，還可去苦澀味，令菜式更美味。
- Refrigerate the beef tenderloin briefly before slicing. The beef will be firmer in texture and easier to slice. The muscle fibres also relax a bit so that the beef will be more tender after cooked.
- When you stir-fry the kale, add some crushed raw cane sugar slab. The sugar will make the kale greener in colour and cover up the bitterness. The dish will taste better overall.

OOKING TIPS

豬扒
pork chop

不敗煎豬扒
Fail-proof pork chop

材料

- 無骨豬扒 3 件
- 雞蛋 1 隻
- 洋葱 1 個（切絲）
- 青、紅椒絲少許

Ingredients

- 3 boneless pork chops
- 1 egg
- 1 onion (shredded)
- red and green chillies (shredded)

醃料

- 生抽 3 湯匙
- 糖 1 茶匙
- 生粉 1 茶匙
- 水 3 湯匙

Marinade

- 3 tbsp light soy sauce
- 1 tsp sugar
- 1 tsp potato starch
- 3 tbsp water

做法

❶ 豬扒洗淨，先以刀背垂直拍鬆一面，翻轉另一面，以橫向拍鬆。

❷ 大碗內，放入醃料拌勻，放入豬扒撈勻，加入已打拂雞蛋拌勻，撲上生粉，備用。

❸ 燒熱油鑊，放入豬扒煎香，加蓋焗煮，可保留豬扒內肉汁；待煎至金黃後開蓋，翻轉另一面，再蓋上鑊蓋以小火焗煮至全熟，盛起後切件，上碟。

❹ 原鑊加入洋葱絲炒勻，下少許鹽及糖調味，放入青、紅椒絲，灒入少許生抽，兜勻後倒入豬扒上即成。

Method

❶ Rinse the pork chops. Tap them with the back of a knife so that the marks of the knife run toward you. Flip them upside down and tap them with the back of a knife so that the marks run from left to right.

❷ Put the marinade into a mixing bowl and mix well. Put the pork chops in. Add a whisked egg. Mix well and coat each pork chop in potato starch.

❸ Heat wok and add oil. Fry the pork chops until golden on one side. Cover the lid when frying to seal in the juices. Flip the pork chops to fry the other side. Cover the lid and fry over low heat until cooked through. Remove and cut into pieces. Save on a serving plate.

❹ In the same wok, put in the shredded onion and toss well. Season with salt and sugar. Add shredded red and green chillies. Sparkle with a dash of light soy sauce. Toss well and pour the mixture over the pork chops. Serve.

必學不敗竅門

- 用刀背拍豬扒，除了可以拍鬆肉質外，還可以防止豬扒因肉質收縮而捲起。
- 煎豬扒時蓋上鑊蓋焗煮片刻，能鎖住肉汁之餘，也能保持肉質嫩滑、彈牙。
- Tapping the pork chops with the back of a knife helps tenderize them. It also prevents the pork chops from curling up after cooked as the muscle fibres contract.
- When you fry the pork chops, cover the lid briefly. It helps sealing in the juices and makes tender and soft.

莧菜
amaranth

莧菜豆腐魚滑羹

Minced dace and tofu thick soup with amaranth

材料

- 莧菜 300 克
- 盒裝蒸煮豆腐 1 盒
- 鯪魚滑 200 克
- 鮮淮山 160 克
- 蛋白 1 隻
- 瑤柱 5 粒
- 上湯 800 毫升
- 杞子少許（浸軟）
- 胡椒粉少許

Ingredients

- 300 g amaranth
- 1 pack tofu for steaming
- 200 g minced dace
- 160 g fresh yam
- 1 egg white
- 5 dried scallops
- 800 ml stock
- dried goji berries (soaked in water till soft)
- ground white pepper

芡汁

- 木薯粉 3-5 茶匙
- 水 45-75 毫升

Glaze

- 3 to 5 tsp tapioca starch
- 45 to 75 ml water

做法

❶ 瑤柱沖洗後浸軟，拆絲備用。
❷ 淮山洗淨、去皮，切粒備用。
❸ 莧菜洗淨，取葉切細，備用。
❹ 豆腐切方粒，備用。
❺ 淮山粒、瑤柱絲放入上湯內，再加入少許油及莧菜煮開，用小匙逐少放入魚滑煮滾，放入豆腐粒，加入芡汁慢慢推至合適濃稠度，最後倒入蛋白稍待凝固（勿攪動），熄火後輕拌勻，盛起，撒上胡椒粉，最後放上杞子裝飾即成。

Method

❶ Rinse the dried scallops and soak them in water till soft. Drain. Break them into fine shreds.
❷ Rinse fresh yam and peel it. Dice and set aside.
❸ Rinse the amaranth. Pick off the leaves and discard the stems. Finely shred the leaves.
❹ Cut the tofu into cubes.
❺ Pour stock into a pot. Add diced yam and dried scallops. Add a dash of oil and amaranth. Bring to the boil. Scoop a small piece of minced dace and put it into the stock. Repeat until all minced dace is used. Add tofu and slowly pour in the glaze while stirring continuously. Check the consistency from time to time as you may not need to use all the glaze. Pour in the egg white at last and stop stirring. Turn off the heat and stir gently. Transfer into serving tureen or bowls. Sprinkled with ground white pepper. Garnish with dried goji berries. Serve.

必學不敗竅門

- 煮湯羹時加入木薯粉，令湯羹更透亮，賣相較佳。
- 享用時，加入麻油提香，令湯羹更滋味。
- When you make thick soup, use tapioca starch as the thickener. It tends to make the soup crystal clear and more attractive.
- Sprinkle with a few drops of sesame oil before serving. It adds an extra aroma to the soup.

購自街市鮮嫩的莧菜。
Young and tender amaranth from wet market.

蝦
shrimp

茄汁煎蝦碌

Fried shrimps in ketchup sauce

材料

- 海中蝦 10 隻
- 洋葱 1/2 個（切粒）
- 薑粒 2 湯匙
- 紅葱頭 4-5 粒（切片）
- 蒜粒 1 湯匙
- 葱花少許

Ingredients

- 10 medium-sized marine shrimps
- 1/2 onion (diced)
- 2 tbsp diced ginger
- 4 to 5 shallots (sliced)
- 1 tbsp diced garlic
- finely chopped spring onion

醬汁

- 茄汁 5 湯匙
- 白醋 1 湯匙
- 糖 1 茶匙

Sauce

- 5 tbsp ketchup
- 1 tbsp white vinegar
- 1 tsp sugar

做法

1. 茄汁、白醋及糖拌勻成醬汁，備用。
2. 蝦洗淨、吸乾水分，剪掉眼、觸鬚、足、頭尾尖角及胃（沙囊），用刀輕切開背去腸，下少許鹽及胡椒粉拌勻，備用。
3. 燒熱油鑊，放入蝦煎香兩面，至 7 成熟，盛起。
4. 燒熱油鑊，先爆香薑粒及紅葱片，放入洋葱粒爆香，再放蒜粒略爆炒，即加入蝦快炒，下少許鹽調味，灒少許紹興酒，放入醬汁炒至汁略收，上碟，灑上葱花裝飾即成。

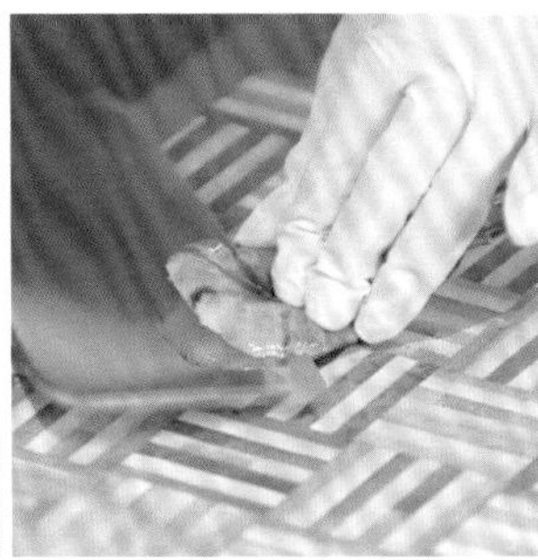

Method

1. Mix all sauce ingredients together until well combined.
2. Rinse the shrimps and wipe dry. Cut off the eyes, rostrums, antennae, feet, pointy tails and the sandy sac with scissors. Then cut the shell along the back. Devein. Add a dash of salt and ground white pepper. Mix well.
3. Heat wok and add oil. Fry the shrimps until both sides golden and about medium-well done. Set aside.
4. Heat wok and add oil. Stir-fry diced ginger and shallots until fragrant. Add diced onion and toss again. Add diced garlic and stir briefly. Put in the shrimps and toss quickly. Season with salt and drizzle with Shaoxing wine. Pour in the sauce from step 1. Toss until the sauce reduces. Save on a serving plate. Garnish with finely chopped spring onion. Serve.

必學不敗竅門

- 茄汁、白醋、糖的份量可因應個人喜愛的酸甜度而增減。
- You may adjust the proportion of ketchup, white vinegar and sugar in the sauce according to your personal preference.

節瓜
Chinese marrow

節瓜蝦米粉絲煲

Chinese marrow with mung bean vermicelli and dried shrimps in clay pot

材料

節瓜 1 個
私房冬菇 5 朵（做法參考 p.14）
蝦米 2-3 湯匙
粉絲 80 克
雞湯 100 毫升
薑 5 片（拍扁）
蒜肉 6-8 瓣（輕拍）

Ingredients

- 1 Chinese marrow
- 5 steamed shiitake mushrooms (method refer to p.14)
- 2 to 3 tbsp dried shrimps
- 80 g mung bean vermicelli
- 100 ml chicken stock
- 5 slices ginger (crushed)
- 6 to 8 cloves garlic (gently crushed)

做法

❶ 冬菇切片；蝦米洗淨、浸軟（水留用）；粉絲沖洗、浸軟。
❷ 節瓜洗淨，用小匙或刀背刮去表皮（盡量保留淺綠部分），切幼條備用。
❸ 熱鑊下油，爆香薑片、蒜肉，加入蝦米爆炒至香味散出，再下節瓜條炒至軟身，加入少許蝦米水、冬菇水及雞湯，下冬菇片及粉絲煮開，按個人口味加入適量鹽調味即成。

Method

❶ Slice the mushroom. Rinse the dried shrimps and soak them in water till soft. Drain and set aside the soaking water. Rinse and soak the mung bean vermicelli in water till soft.
❷ Rinse the Chinese marrow. Scrape off the dark green skin with a metal spoon or the back of a knife. (It's okay to have a bit light green on. It doesn't have to look completely white.) Cut into fine strips. Set aside.
❸ Heat wok and add oil. Stir-fry ginger and garlic cloves until fragrant. Put in the dried shrimps and stir-fry until fragrant. Add Chinese marrow and toss until it softens. Add the soaking water from dried shrimps and shiitake mushrooms. Pour in the chicken stock. Add sliced mushrooms and mung bean vermicelli. Bring to the boil. Season with salt according to your taste. Serve.

必學不敗竅門

- 放入蝦米水、冬菇水煮節瓜，令菜式更原汁原味。
- 節瓜條要先炒至軟身才加入粉絲，否則會因煮熟的時間不同，而令粉絲糊掉。
- Cook the Chinese marrow in water used to soak dried shrimps and shiitake mushrooms. This step would give this dish extra depth of flavours.
- Stir-fry the Chinese marrow strips until soft before adding the mung bean vermicelli. As the mung bean vermicelli literally take seconds to cook, putting them in too early and they will turn soggy.

排骨

pork rib

豉汁蒸排骨

Steamed pork ribs in black bean sauce

材料

- 排骨 600 克（切粒）
- 蒜肉 6-8 瓣

Ingredients

- 600 g pork ribs (diced)
- 6 to 8 cloves garlic

醃料

- 生抽 1 湯匙
- 糖 3/4 茶匙
- 麻油 1 湯匙
- 生粉 1 茶匙

Marinade

- 1 tbsp light soy sauce
- 3/4 tsp sugar
- 1 tbsp sesame oil
- 1 tsp potato starch

豉汁醬

- 豆豉 2 湯匙
- 糖 1/2 茶匙
- 紹興酒 1 湯匙
- 蒜粒 1.5 湯匙
- 指天椒 1 隻（切圈）

Black bean sauce

- 2 tbsp fermented black beans
- 1/2 tsp sugar
- 1 tbsp Shaoxing wine
- 1.5 tbsp diced garlic
- 1 bird's eye chilli (cut into rings)

做法

❶ 排骨加入生抽及糖拌勻，再加入麻油及生粉略醃。
❷ 蒜肉先放上蒸碟內，蒸 3-5 分鐘，備用。
❸ 處理豉汁醬：豆豉沖洗乾淨，吸乾後用湯匙壓至半爛，加入糖、紹興酒、蒜粒及指天椒拌勻。
❹ 豉汁醬與排骨拌勻，加水 2-3 湯匙拌勻，放進蒜肉蒸碟內，均勻地鋪好，放入蒸爐蒸約 20 分鐘（時間視火力調節而定），或大火蒸 12-15 分鐘即成。

Method

❶ To marinate the pork ribs, add light soy sauce and sugar to the pork ribs first. Mix well. Then add sesame oil and potato starch. Mix well and leave them briefly.
❷ Put the garlic cloves on the steaming plate. Steam for 3 to 5 minutes. Set aside.
❸ To make the black bean sauce, rinse the fermented black beans. Wipe dry and crush them with a tablespoon until coarsely mashed. Add sugar, Shaoxing wine, diced garlic and bird's eye chilli. Mix well.
❹ Mix the pork ribs with the black bean sauce. Add 2 to 3 tbsp of water and stir again. Transfer the mixture into the steaming plate on top of the garlic cloves. Spread the mixture evenly so that no two pieces of pork stack on top of each other. Steam in an electric steamer for 20 minutes (adjust the steaming time according to the power of your steamer). Or steam in a wok filled with water over high heat for 12 to 15 minutes. Serve.

必學不敗竅門

- 美儀教大家，排骨宜選肥瘦比例為 4 比 6 的肉排，配搭豉汁份外滋味。
- According to Elena, the best pork ribs for this recipe should have 40% of fat to 60% of lean meat. They work best with the black bean sauce.

金沙骨
pork sparerib

酸梅炆排骨

Braised pork ribs with sour plums

材料

- 金沙骨 600 克
- 冰花梅醬 2 湯匙
- 酸梅 3 粒
- 醃酸薑 3 塊
- 上湯 30 毫升
- 薑 3 片（拍扁）
- 紅尖椒少許（切角）

Ingredients

- 600 g pork spareribs
- 2 tbsp Chinese plum sauce
- 3 sour pickled plums
- 3 pieces pickled ginger
- 30 ml stock
- 3 slices ginger (crushed)
- red chillies (cut into wedges)

醃料

- 鹽 1/4 茶匙
- 生抽 1 湯匙
- 糖少許
- 生粉 1 茶匙
- 紹興酒 2 茶匙
- 麻油少許

Marinade

- 1/4 tsp salt
- 1 tbsp light soy sauce
- sugar
- 1 tsp potato starch
- 2 tsp Shaoxing wine
- sesame oil

做法

1. 金沙骨洗淨，以鹽醃勻，烹調前加入餘下的醃料拌勻，備用。
2. 酸梅用刀背剁爛，備用。
3. 燒熱油鑊，先爆香薑片，加入金沙骨（骨朝上），用中火煎至每面微焦香，至 6-7 成熟，下酸梅蓉及冰花梅醬炒勻，再加入上湯煮滾，下糖 1 茶匙調味，加入醃酸薑，加蓋，轉小火炆煮約 6-8 分鐘至醬汁濃稠，最後放入紅尖椒炒勻，上碟即成。

Method

1. Rinse the spareribs. Sprinkle 1/4 tsp of salt over them and rub evenly. Leave them briefly. Add the rest of the marinade right before cooking and mix well.
2. Mash the sour pickled plums by tapping them with the back of a knife.
3. Heat wok and add oil. Stir-fry ginger until fragrant. Put in the spareribs with the bone side up. Fry over medium heat until lightly browned on all sides and about medium-well done. Add sour pickled plums and plum sauce. Toss well. Add stock and bring to the boil. Add 1 tsp of sugar and pickled ginger. Cover the lid and turn to low heat. Simmer for about 6 to 8 minutes until the sauce reduces. Add red chillies at last. Toss and serve.

必學不敗竅門

- 排骨先煎香表面至金黃色，可鎖住肉汁。
- 如不嗜酸者，可多加點糖中和酸梅的味道。
- 加入醃酸薑同煮，酸酸甜甜的薑味，提升味道層次。
- Fry the ribs until golden on all sides to seal in the juices.
- If you're not a big fan of sour dishes, you may use more sugar to balance out the acidity of the plums.
- The pickled ginger lends a sweet-and-sour zing to the dish, taking the flavours to another dimension.

COOKING TIPS

大魚頭
head of bighead carp

馬蹄沙魚雲羹
Fish head thick soup with water chestnut

材料

- 大魚頭 1 個
- 馬蹄 10 粒
- 魚上湯 400 毫升
- 雪蓮子 2 湯匙
- 薑絲 1 湯匙
- 果皮 1 瓣
- 馬蹄粉 3 茶匙
- 木薯粉 1/2 茶匙
- 芹菜少量（切粒）
- 紅尖椒少量（切絲）
- 芫茜少量

Ingredients

- 1 head of bighead carp
- 10 water chestnuts
- 400 ml fish stock
- 2 tbsp dried honeylocust seeds
- 1 tbsp shredded ginger
- 1 piece dried tangerine peel
- 3 tsp water chestnut starch
- 1/2 tsp tapioca starch
- Chinese celery (diced)
- red chillies (shredded)
- coriander

示範短片

做法

❶ 馬蹄洗淨、去皮，用刀背拍碎備用。
❷ 雪蓮子洗淨，清水浸過夜，隔水蒸 5 分鐘備用。
❸ 果皮洗淨、浸軟，切幼絲，備用。
❹ 預備馬蹄芡：馬蹄粉先用湯匙壓碎後過篩，混入木薯粉，加適量水拌勻備用。
❺ 大魚頭洗淨，大火隔水蒸 15 分鐘。蒸完後碟內魚湯汁留用，取出魚頭放大碗內，加冷開水蓋過面，用手在水中拆肉去骨，完成後過篩隔走魚骨，水留用。
❻ 把一半魚水、蒸魚剩下魚湯汁及魚上湯煮滾，放進馬蹄碎、雪蓮子煮開，加入果皮絲、薑絲、魚雲及魚肉煮沸，放進適量鹽、糖調味，逐少加入馬蹄芡煮至合適狀，放魚露 1 茶匙提鮮，盛起。
❼ 最後撒上適量胡椒粉，綴以芹菜粒、芫茜、紅椒絲即成。

Method

1. Rinse the water chestnuts and peel them. Crush into small bits with the back of a knife. Set aside.
2. Rinse the honeylocust seeds. Soak in water overnight. Steam for 5 minutes. Drain and set aside.
3. Rinse the dried tangerine peel and soak in water till soft. Finely shred it.
4. To make the water chestnut starch thickening glaze, put water chestnut starch into a bowl and crush with a tablespoon. Sift it and add tapioca starch. Add some water and mix well into a smooth glaze.
5. Rinse the fish head and put on a steaming dish. Steam over high heat for 15 minutes. Drain any juices in the dish and save for later use. Transfer the fish head into a big bowl. Add cold drinking water to cover. Debone the fish head under water. Remove as many bones as you can with your hands. Set aside the deboned fish head. Then strain the water to remove remaining bones. Save the water for later use.
6. Pour half of the water from deboning the fish head into a pot. Add the juices from steamed fish head and fish stock. Bring to the boil. Add water chestnuts and honeylocust seeds. Bring to the boil again. Add dried tangerine peel, shredded ginger, and de-boned fish head. Bring to the boil. Season with salt and sugar. Slowly stir in the water chestnut thickening glaze from step 4 to desired consistency. Add 1 tsp of fish sauce for extra flavours. Save in a tureen or serving bowls.
7. Sprinkle with ground white pepper. Garnish with diced Chinese celery, coriander and shredded red chillies. Serve.

必學不敗竅門

- 蒸後的魚頭很熱，為免燙手和易於拆肉，將魚雲放入凍滾水內拆肉，魚骨會沉底，而魚肉會浮起；隔渣後的水可用來煮湯。
- 如時間不足來不及將馬蹄粉壓碎後過篩，可用清水將馬蹄粉調勻至沒有顆粒才放入木薯粉。
- The fish head is very hot after steamed. To cool it down and to make it easier to debone, it's advisable to put the fish head in cold drinking water. Any loose bones will sink to the bottom and the flesh and connective tissue will float. You can strain the water and use it as a base for the soup.
- If you run out of time, you don't have to crush the water chestnut starch and sift it. Just add water to the water chestnut starch and stir until lump-free before adding tapioca starch.

魚肚
fish tripe

陳皮洋薏米肉碎魚肚羹

Fish tripe thick soup with aged tangerine peel, pearl barley and ground pork

材料

- 脢頭肉 200 克
- 洋薏米 40 克
- 砂爆魚肚 1 件（約 40 克）
- 無糖豆腐花 200 克
- 粥水 500 毫升
- 大地魚 1/6 條
- 陳皮 1 瓣
- 魚露 2 茶匙
- 薑 2-3 片（切絲）
- 芹菜粒少許
- 胡椒粉適量

Ingredients

- 200 g pork shoulder butt
- 40 g pearl barley
- 1 piece puffed fish tripe (about 40 g)
- 200 g unsweetened tofu pudding
- 500 ml rice gruel (strained, liquid only)
- 1/6 dried plaice
- 1 piece aged tangerine peel
- 2 tsp fish sauce
- 2 to 3 slices ginger (finely shredded)
- diced Chinese celery
- ground white pepper

做法

1. 脢頭肉洗淨，加入粗鹽 1-2 茶匙醃勻，放雪櫃冷藏一晚成鹹肉，取出，抹去粗鹽，剁碎備用。
2. 洋薏米洗淨，浸泡 1-2 小時；陳皮浸軟，切絲備用。
3. 砂爆魚肚洗淨，用水浸泡至軟身，加入鹽少許、油約 2 茶匙及生粉 2 茶匙塗勻，蒸 5 分鐘，切粗粒備用。
4. 大地魚原條以明火燒香，燒至邊及骨位微微焦燶，用小刀刮掉黑色焦，取 1/6 剪成大丁方（其他留用），備用。
5. 粥水放鍋內煮開，取少許與肉碎拌勻，放回鍋內，加入魚肚及洋薏米，邊攪拌邊煮滾。
6. 最後加入陳皮絲、魚露、無糖豆腐花、薑絲，逐少加入生粉水埋芡，輕拌勻，盛起。
7. 放上大地魚及芹菜粒，撒上胡椒粉即成。

Method

1. Rinse the pork. Rub 1 to 2 tsp of coarse salt evenly on it. Refrigerate overnight. Remove any salt on the surface. Rinse well. Finely chop the pork. Set aside.
2. Rinse the pearl barley. Soak in water for 1 to 2 hours. Drain and set aside. Soak aged tangerine peel in water till soft. Finely shred it. Set aside.
3. Rinse the puffed fish tripe. Soak in water till soft. Drain and squeeze dry. Add a pinch of salt, 2 tsp of oil and 2 tsp of potato starch. Mix well. Steam for 5 minutes. Dice coarsely. Set aside.
4. Grill the whole dried plaice over fire until lightly browned on the edges and on the bones. Scrape off any burnt bits with a knife. Cut about 1/6 of the fish for this recipe and use the rest for other recipes. Cut into large squares. Set aside.
5. Pour the rice gruel into a pot. Bring to the boil. Pour some into the chopped pork from step 1. Mix well and pour the mixture back into the pot. Add fish tripe and pearl barley. Keep stirring while heating it up.
6. Add aged tangerine peel, fish sauce and unsweetened tofu pudding at last. Stir in potato starch thickening glaze slowly while stirring gently. Save in a tureen or serving bowls.
7. Arrange dried plaice squares and diced Chinese celery on top. Sprinkle with ground white pepper. Serve.

必學不敗竅門

- 大地魚烘後才帶香味，烘至邊位微焦，再刮走燒焦部分，以免吃時有焦燶苦味。
- 加入豆腐花令湯羹更滑溜。
- Dried plaice has to be grilled to bring out the flavours. Just burn over naked flame until the edges look burnt. Then scrape off the burnt bits. Otherwise, the soup may taste bitter.
- Adding tofu pudding make the soup velvety in texture.

紅衫魚
golden thread fish

冬瓜魚蓉雪燕粥

Congee with winter melon, golden thread fish and tragacanth resin

必學不敗竅門

- 想將冬瓜煮成燕窩的口感，秘訣在於切絲後蘸上生粉，放入滾水煮一會，熄火焗至透明。
- 米先與皮蛋拌勻，令煲出來的粥更綿滑。
- The shredded winter melon shares the same texture as bird's nest in this recipe. Just make sure you coat it in potato starch after shredding it. Then blanch it in boiling water briefly and cover the lid. Let the remaining heat cook through the winter melon until transparent.
- Mashing the thousand-year egg and mixing it with rice before adding water to make congee. The congee will be creamier this way.

COOKING TIP

材料

- 紅衫魚 2-3 條（細條）
- 冬瓜 100 克
- 雪燕 2 小粒
- 米 1 量米杯
- 皮蛋 1/4 隻
- 大頭菜 1/2 片
- 薑絲適量
- 葱絲適量
- 果皮 1 瓣

Ingredients

- 2 to 3 small golden thread fish
- 100 g winter melon
- 2 small pieces tragacanth resin
- 1 cup rice (measured with the cup that comes with the rice cooker)
- 1/4 thousand-year egg
- 1/2 slice salted Cantonese kohlrabi
- shredded ginger
- finely chopped spring onion
- 1 piece dried tangerine peel

做法

1. 雪燕洗淨，浸過夜，瀝乾水分，放入沸水氽燙一會，盛起備用。
2. 果皮洗淨，浸軟後切幼絲；大頭菜洗淨，浸泡，切細粒。
3. 冬瓜洗淨，去皮，切幼絲，加少許生粉拌勻，放入熱水以小火煮 5 分鐘，熄火，加蓋焗至透明。
4. 米洗淨，混合皮蛋壓勻，放入水內煮成粥底。
5. 紅衫魚洗淨，下油鑊煎至金黃，取出起肉，魚骨、魚頭沖入熱水煮成魚湯，備用。
6. 把適量魚湯加入粥底調至適合的濃稠度，加入魚肉、雪燕、果皮絲，下少許鹽調味，最後下冬瓜絲拌勻，盛起，放上薑絲、葱絲、大頭菜粒即成。

Method

1. Rinse the tragacanth resin. Soak it in water overnight. Drain and blanch in boiling water briefly. Drain and set aside.
2. Rinse the dried tangerine peel. Soak in water till soft. Shred it. Set aside. Rinse the salted kohlrabi. Soak in water for a while. Drain and dice finely.
3. Rinse winter melon. Peel it and shred finely. Add a pinch of potato starch. Mix well. Blanch in boiling water over low heat for 5 minutes. Cover the lid and turn off the heat. Leave the winter melon in the water until it turns transparent.
4. Rinse and drain the rice. Add the thousand-year egg. Crush the egg and mix well with the rice. Add water and bring to the boil in a pot. Cook until the rice breaks down and the mixture turns creamy and thick.
5. Rinse the fish. Fry in a wok in some oil until golden on both sides. Debone the fish. Set aside the flesh. Put the fish heads and bones back in the wok. Pour in boiling water. Cook for a while to make the fish stock.
6. Pour some of the fish stock from step 5 into the congee from step 4. Adjust the amount of fish stock used according to your preferred consistency. Add fish flesh, tragacanth resin and dried tangerine peel. Season with a pinch of salt. Add winter melon and mix well. Save in serving bowls or a tureen. Arrange shredded ginger, spring onion and diced salted kohlrabi on top. Serve.

石崇魚
scorpion fish

無添加奶白魚湯配銀針粉

Silver needle noodles in all-natural milky fish stock

材料

- 石崇魚 600 克
- 金華火腿 60 克
- 雪燕 2 小粒
- 勝瓜 1/4 條（去皮、切粒）
- 冬瓜 100 克
- 大地魚粉 1/2 茶匙
- 銀針粉 200 克（幼身）
- 洋葱 1 個（切絲）
- 薑 4-5 片（拍扁）

Ingredients

- 600 g scorpion fish
- 60 g Jinhua ham
- 2 small pieces tragacanth resin
- 1/4 angled loofah (peeled and diced)
- 100 g winter melon
- 1/2 tsp ground dried plaice
- 200 g silver needle noodles (finer variety)
- 1 onion (shredded)
- 4 to 5 slices ginger (crushed)

做法

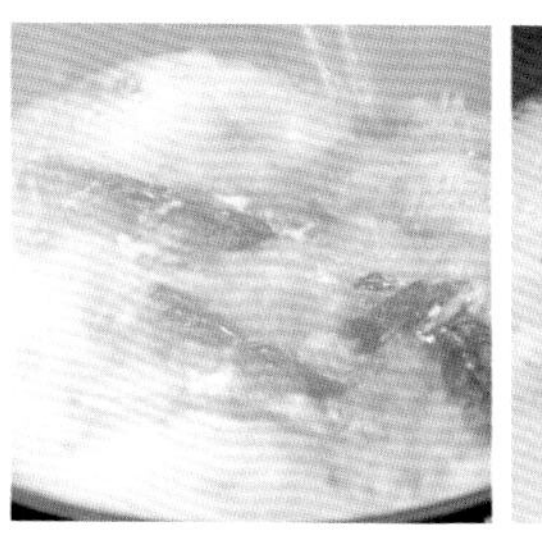

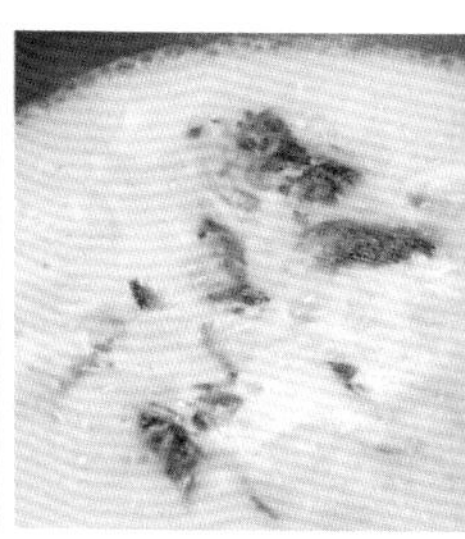

❶ 雪燕用清水浸過夜，隔水蒸 5 分鐘備用。

❷ 金華火腿洗淨，浸熱水洗走多餘油分，切細粒備用。

❸ 冬瓜洗淨，去皮，切粒，放進熱水焗熟至透明，備用。

❹ 銀針粉用熱水煮開，盛起備用。

❺ 石崇魚洗淨，吸乾水分。熱鑊下油，爆香薑片，下石崇魚以中小火煎至兩面金黃。以廚房紙抹走鑊內多餘油分，沖入熱水，加入洋葱絲，加蓋煮約 10 分鐘至奶白色，用隔篩濾去魚骨及洋葱絲，魚湯回鑊煮滾。

❻ 冬瓜粒、金華火腿放入魚湯內，加入大地魚粉、鹽、胡椒粉各適量調味，下雪燕、勝瓜粒、銀針粉煮滾即成。

Method

❶ Soak tragacanth resin in water overnight. Drain and steam for 5 minutes. Set aside.

❷ Rinse Jinhua ham. Soak in hot water to remove the stale grease. Finely dice it.

❸ Rinse winter melon and peel it. Dice and blanch in boiling water. Cover the lid and leave it in the water until it turns transparent. Drain.

❹ Boil silver needle noodles in water. Drain.

❺ Rinse the scorpion fish. Wipe dry. Heat oil in wok. Stir-fry ginger until fragrant. Put in the fish and fry over medium-low heat until both sides golden. Drain any oil in the wok and wipe dry with paper towel. Pour in boiling water. Add shredded onion. Cover the lid and cook for about 10 minutes until the water turns milky. Strain the soup to remove the fish bones and onion. Pour the soup back in the wok. Bring to the boil.

❻ Add winter melon and Jinhua ham in the soup. Add ground dried plaice, salt and ground white pepper. Finally put in tragacanth resin, angled loofah and silver needle noodles. Bring to the boil. Serve.

必學不敗竅門

- 鑊的熱度必須足夠，煎魚後潑入沸水，可成功煲成奶白色魚湯。
- 煎魚後用廚房紙抹去多餘油分，減少魚湯的油膩。
- When you fry the fish, make sure the wok is smoking hot. Then pour in boiling water after the fish is fried on both sides. That is the key to making milky white fish stock.
- After frying the fish, it is advisable to drain and wipe off any oil in the wok. That would make the fish stock less greasy.

COOKING TIPS

海鮮

seafood

豉油皇海鮮炒麵

Seafood fried noodles with soy sauce

做法

1. 鮮魷洗淨、去外皮，吸乾水分，切成圈。
2. 帶子解凍，洗淨，吸乾水分，橫切一半，撲上少許生粉，備用。
3. 全蛋麵放於滾水充分弄散，立即取出，放入冷水浸泡降溫，盛起，瀝乾水分。
4. 生抽、鮮醬油，老抽及同等份量的水拌勻，加入大地魚粉調味，備用。
5. 熱鑊下油，大火炒香洋葱絲及銀芽至半熟，盛起。
6. 熱鑊下油，帶子煎至半熟，盛起，熄火。以餘溫略炒鮮魷至半熟，盛起備用。
7. 熱鑊下油，放入全蛋麵煎烘至兩面香脆微焦，下調味料並用筷子快速撥散，加入洋葱、銀芽、帶子及鮮魷快炒，最後放入葱段及紅椒絲炒勻，上碟後撒上芝麻即成。

必學不敗竅門

- 炒麵要好看，烘麵前必須弄散麵條，而且慢慢烘至邊緣微焦，才加入其他材料拌炒。
- 老抽先與水調勻才加入調味，顏色會更均勻。
- For the best presentation, always separate the noodles first before frying them in oil. When you fry them, do it slowly until the edges are lightly browned first, before adding other ingredients.
- Mixing the dark soy sauce with some water would help the noodles pick up the colour more evenly.

COOKING TIP

材料	Ingredients
全蛋麵 1 個	• 1 nest dried egg noodles for stir-frying
鮮魷 1 隻	• 1 squid
帶子 6 粒	• 6 scallops
洋葱 1/2 個（切絲）	• 1/2 onion (shredded)
銀芽 80 克	• 80 g mung bean sprouts
葱 3 條（切段）	• 3 spring onion (cut into short lengths)
紅椒絲少許	• red chillies (finely shredded)
炒香芝麻適量	• toasted sesames

調味料	Seasoning
生抽 2 茶匙	• 2 tsp light soy sauce
鮮醬油 1 茶匙	• 1 tsp Maggi seasoning
老抽 2 湯匙	• 2 tbsp dark soy sauce
大地魚粉 1 茶匙	• 1 tsp ground dried plaice

Method

❶ Rinse the squid. Peel off the purple skin. Wipe dry. Cut into rings.

❷ Thaw the scallops. Rinse and wipe dry. Slice each scallop in half crosswise. Coat them lightly in potato starch. Set aside.

❸ Blanch the egg noodles in boiling water until they scatter. Drain and dunk into ice water immediately to cool them. Drain well.

❹ To make the seasoning, put light soy sauce, Maggi seasoning and dark soy sauce into a bowl. Mix well and add equal volume of water. Mix again. Add ground dried plaice.

❺ Heat wok and add oil. Stir-fry onion and mung bean sprouts over high heat until onion is fragrant and bean sprouts are half-cooked. Set aside.

❻ Heat wok and add oil. Fry the scallops until half cooked. Set aside and turn off the heat. Put in the squid and stir-fry in the residual heat until the squid is half-cooked. Set aside.

❼ Heat wok and add oil. Put in the egg noodles. Fry until both sides lightly browned and crispy. Add the seasoning mixture. Toss with a pair of chopsticks quickly. Add onion, mung bean sprouts, scallops and squid. Stir quickly. Put in spring onion and red chillies at last. Toss again. Transfer onto a serving plate. Sprinkle with toasted sesames. Serve.

白鱈魚
white cod fillet

豆酥鱈魚

Fried cod fillet cubes with fried savory crisbean

材料

- 法國白鱈魚 400 克
- 豆酥 100 克
- 蒜粒 3 湯匙
- 豆瓣醬 1 湯匙
- 鹹蛋黃半隻
- 無鹽牛油 20 克
- 紹興酒少許

Ingredients

- 400 g French white cod fillet
- 100 g fried savory crisbean
- 3 tbsp diced garlic
- 1 tbsp spicy bean sauce
- 1/2 salted egg yolk
- 20 g unsalted butter
- Shaoxing wine

醃料

- 鹽、胡椒粉各少許
- 生粉 2 茶匙

Marinade

- salt
- ground white pepper
- 2 tsp potato starch

做法

❶ 鹹蛋黃蒸熟，切碎備用。
❷ 白鱈魚解凍，洗淨，切大粒，塗上醃料略醃。
❸ 熱鑊下油，下白鱈魚粒煎至金黃色。
❹ 另一熱油鍋，先爆香蒜粒及豆瓣醬，下鹹蛋黃碎略爆炒，加入豆酥不停翻炒，下牛油繼續炒勻（如太乾可酌加煮食油），最後灒入紹興酒炒勻，將 2/3 豆酥料放於碟上。
❺ 將煎好的白鱈魚粒放豆酥上，撒入餘下豆酥即成。

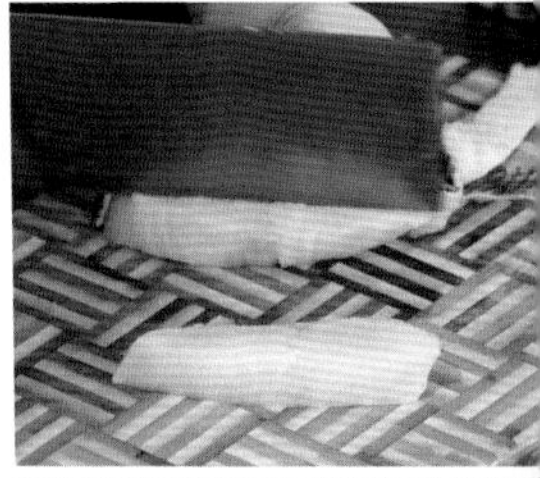

Method

❶ Steam the salted egg yolk until cooked. Chop finely.
❷ Thaw the cod. Rinse and dice coarsely. Add marinade and mix well.
❸ Heat wok and add oil. Fry the diced cod until golden.
❹ In another wok, stir-fry garlic and spicy bean sauce until fragrant. Add salted egg yolk. Toss briefly. Put in the fried savory crisbean and keep on tossing. Add butter and toss further. If it looks dry, add some oil at this point. Drizzle with Shaoxing wine and toss again. Transfer 2/3 of this mixture on a serving plate.
❺ Arrange the fried cod cubes over the fried savory crisbean mixture. Then sprinkle the rest of the fried savory crisbean mixture over. Serve.

必學不敗竅門

- 建議用慢火將鱈魚煎熟至金黃色。
- 豆酥油香十足，在雜貨店有售。加入鹹蛋炒勻，味道濃郁，毋須另加調味料。
- Fry the cod over low heat until cooked through and golden.
- Fried savory crisbean is flavourful and aromatic. You can get it from conventional grocery stores. The savory crisbean will pick up the rich flavours of the salted egg yolk and you don't need to season it any further.

COOKING TIPS

雞扒

chicken thigh

香橙雞柳

Fried chicken strips in orange custard sauce

必學不敗竅門

- 切雞扒時皮向下，不會滑開，容易切成塊狀。
- 在炸漿內加入少許吉士粉，炸出來的雞扒色澤金黃，口感鬆脆。
- When you slice the chicken thigh, put it on a chopping board with the skin side down. It won't slide on the chopping board and it's easier to slice that way.
- Adding some custard powder into the deep-frying batter would accentuate the yellow colour of the egg yolk. The chicken strips would look more golden with the crispy crust.

COOKING TIP

材料

- 雞扒 2 件
- 蛋黃 1 隻
- 生粉 3 湯匙
- 橙 1 個
- 腰果 3 粒（搗碎）
- 炒香芝麻少許
- 炸碧綠枸杞葉適量
- 橙皮碎適量

Ingredients

- 2 boneless chicken thighs
- 1 egg yolk
- 3 tbsp potato starch
- 1 orange
- 3 cashew nuts (crushed)
- toasted sesames
- deep-fried goji leaves
- grated orange zest

香橙醬汁

- 鮮橙汁 200 毫升
- 芒果汁 3 湯匙（連果蓉）
- 白醋 1 茶匙
- 吉士粉 1 茶匙
- 糖 1.5 茶匙
- 橙 1 個（刮皮用）

Orange custard sauce

- 200 ml freshly squeezed orange juice
- 3 tbsp mango puree
- 1 tsp white vinegar
- 1 tsp custard powder
- 1.5 tsp sugar
- 1 orange (for grated zest)

做法

❶ 雞扒洗淨，以少許鹽、胡椒粉略醃，切成雞柳，加入蛋黃及生粉拌勻，備用。

❷ 橙洗淨，部分起肉、部分榨汁；橙皮（只取橙色最外部分）切絲。

❸ 取橙汁與芒果汁混和，加入白醋拌勻，備用。

❹ 吉士粉加 3.5 茶匙水調稀，備用。

❺ 煮香橙醬汁：熱鑊下水 3 湯匙，下糖煮溶，加入橙皮絲，依次放入橙芒汁及吉士粉水，煮成濃稠香橙醬汁。

❻ 熱鑊下油，下雞柳半煎炸至金黃熟透，盛起備用。

❼ 熱鑊下少許油，放入橙肉及雞柳，加入適量香橙醬汁拌勻上碟，最後撒上腰果碎及芝麻，以炸枸杞葉及橙皮碎裝飾即成。

Method

❶ Rinse the chicken thighs. Sprinkle with salt and ground white pepper. Mix well and leave it briefly. Cut into strips. Add egg yolk and potato starch. Mix well. Set aside.

❷ Rinse the orange. Grate the zest of the orange and set aside. (Or, scrape off the orange peel with a knife and shred it. Don't use the white skin.) Then cut the orange in half. Squeeze the juice of half an orange. Then segment the other half.

❸ Mix orange juice with mango puree. Add white vinegar. Mix again.

❹ Add 3.5 tsp of water to the custard powder. Mix well.

❺ To make the orange custard sauce, heat a wok and add 3 tbsp of water. Add sugar. Cook until sugar dissolves. Put in the orange zest and then add the orange juice mixture from step 3. Put in the custard powder mixture from step 4. Stir well. Cook until it thickens.

❻ Heat wok and add oil. Put in the chicken strips and cook in semi-deep frying manner until golden and cooked through. Drain and set aside.

❼ Heat wok and add oil. Put in the orange segments and the fried chicken strips. Pour in the orange custard sauce from step 5. Toss to coat well. Save on a serving plate. Sprinkle with cashew and sesames on top. Garnish with deep-fried goji leaves and orange peel.

豆豉雞煲

Black bean chicken in clay pot

材料

- 雞 1 隻
- 紅葱頭 6 粒
- 薑 2-3 片（拍扁）
- 蒜肉 4-5 瓣（拍扁、切粒）
- 指天椒 1-2 隻（切圈）
- 豆豉（原粒）1 茶匙
- 玫瑰露 1 湯匙
- 生粉芡適量
- 葱段適量

Ingredients

- 1 chicken
- 6 shallots
- 2 to 3 slices ginger (crushed)
- 4 to 5 cloves garlic (crushed and diced)
- 1 to 2 bird's eye chillies (cut into rings)
- 1 tsp whole fermented black beans
- 1 tbsp Chinese rose wine
- potato starch thickening glaze
- spring onion (cut into short lengths)

醃料

- 生抽 1/2 湯匙
- 老抽 1 湯匙
- 糖 1/2 茶匙
- 生粉 1/2 湯匙
- 麻油 1 湯匙

Marinade

- 1/2 tbsp light soy sauce
- 1 tbsp dark soy sauce
- 1/2 tsp sugar
- 1/2 tbsp potato starch
- 1 tbsp sesame oil

豆豉醬

- 豆豉 1 湯匙
- 糖 1 茶匙
- 蒜蓉 2 湯匙
- 玫瑰露 1/2 湯匙

Black bean sauce

- 1 tbsp fermented black beans
- 1 tsp sugar
- 2 tbsp grated garlic
- 1/2 tbsp Chinese rose wine

煮雞汁

- 生抽 2 湯匙
- 老抽 2 湯匙
- 糖 2 湯匙
- 蠔油 2 湯匙
- 水 1 湯匙

Seasoning

- 2 tbsp light soy sauce
- 2 tbsp dark soy sauce
- 2 tbsp sugar
- 2 tbsp oyster sauce
- 1 tbsp water

- 用玫瑰露煮雞，兩者是絕配，能夠增加香味，提升層次。
- Chinese rose wine is the perfect match for chicken dishes. It elevate the aromas and flavours to another level.

COOKING TIP

做法

❶ 雞洗淨、斬件，以醃料醃 3 小時以上（醃過夜更佳），備用。
❷ 紅葱頭切半、拍鬆，備用。
❸ 準備豆豉醬：豆豉先壓爛，加入其餘材料拌勻，待用。
❹ 將所有煮雞汁材料拌勻，待用。
❺ 熱鑊下油，加入已吸乾的雞件煎香至金黃色，盛起備用。
❻ 燒熱砂鍋，下油，放入薑片及紅葱頭爆香，再下蒜肉爆香，加入豆豉醬及雞件炒勻，倒入煮雞汁及原粒豆豉，炒勻後加蓋燜煮數分鐘至水分收乾。
❼ 放入指天椒，灒玫瑰露，下生粉芡，最後放入葱段，加蓋稍焗一會即成。

Method

❶ Rinse the chicken and chop into pieces. Add marinade and mix well. Leave it for at least 3 hours (or overnight preferably).
❷ Cut shallots in half. Crush them.
❸ To make the black bean sauce, crush the black beans. Add all remaining black bean sauce ingredients. Mix well.
❹ Mix all seasoning together.
❺ Heat wok and add oil. Drain the marinated chicken pieces and wipe dry with paper towel. Fry the chicken pieces in the wok until golden. Drain and set aside.
❻ Heat a clay pot. Add oil and stir-fry ginger and shallots until fragrant. Add garlic cloves and stir again. Add black bean sauce and chicken pieces. Toss well. Pour in the seasoning mixture and whole fermented black beans. Toss to mix well. Cover the lid and cook for a few minutes until the sauce reduces. Turn off the heat.
❼ Add bird's eye chillies. Drizzle with Chinese rose wine. Stir in the potato starch thickening glaze. Top with spring onion. Cover the lid and leave it briefly.

chicken

爸爸的冬菇雞飯

Dad's clay pot rice with chicken and shiitake mushrooms

材料

- 鮮雞 1/2 隻
- 私房冬菇 8 朵（做法參考 p.14）
- 米 2.5 量米杯
- 生粉 2 湯匙
- 煲仔飯甜豉油 3 湯匙

醃料

- 豉油雞汁 4 湯匙
- 薑汁 2 湯匙
- 玫瑰露 1 湯匙

Ingredients

- 1/2 freshly slaughtered chicken
- 8 steamed shiitake mushrooms (method refer to p.14)
- 2.5 cups rice (measured with the cup that comes with the rice cooker)
- 2 tbsp potato starch
- 3 tbsp sweet soy sauce for clay pot rice

Marinade for chicken

- 4 tbsp soy marinade for chicken
- 2 tbsp ginger juice
- 1 tbsp Chinese rose wine

做法

❶ 雞洗淨，斬件，加醃料拌勻，放雪櫃醃 3 小時或過夜。
❷ 米洗淨，放砂鍋內，加入等量的冬菇水，大火煮滾，再轉細火慢煮。
❸ 雞件撲上生粉，放熱油鍋內煎至金黃半熟，盛起備用。
❹ 待飯煮至水剛收乾，把雞件及冬菇鋪上飯面，蓋好焗煮多一會，期間不停轉動鍋子令整個鍋邊均勻受熱，待陣陣飯焦香飄出即可熄火。
❺ 最後淋上甜豉油，開火及轉動鍋子，待甜豉油受熱均勻即熄火。將飯粒拌勻，撒上葱段即成。

Method

❶ Rinse the chicken and chop into pieces. Add marinade and mix well. Refrigerate for 3 hours or overnight.
❷ Rinse the rice and put it in a clay pot. Add the same volume of water for soaking shiitake mushrooms. Bring to the boil over high heat. Turn to low heat and cover the lid.
❸ Coat the chicken pieces in potato starch. Fry in oil until golden and half-cooked. Drain and set aside.
❹ Cook the rice until there is no longer liquid over the rice. Arrange the chicken and shiitake mushrooms over the rice. Cover the lid and cook for a while longer. In the meantime, tip and turn the pot so that all sides of the pot receive the same amount of heat. When you can smell the rice burning slightly, you can turn off the heat.
❺ Drizzle with sweet soy sauce at last. Turn on the heat and turn the pot. Turn off the heat when the soy sauce is heated through. Fluff up the rice and sprinkle with spring onion. Serve.

必學不敗竅門

- 雞件先用油略煎，能鎖住肉汁，毋須煎至全熟。
- 炮製煲仔飯最重要的步驟是最後熄火焗一會，能令飯粒更好吃。
- Par-frying the chicken helps seal in the juices. You don't need to cook it through at this stage.
- For clay pot rice, it's important to leave the rice in the pot for a while after turning off the heat. The residual heat would steam the rice and make it taste better.

番茄
tomato

不經典番茄炒蛋
The unorthodox scrambled egg with tomato

材料

番茄 5 個
雞蛋 6 隻
雞柳 200 克
蒜肉 4-6 瓣（切粒）
上湯 3-4 湯匙
葱花 1-2 茶匙
炒香芝麻少許

調味料

糖 1.5 茶匙
鹽 1 茶匙

Ingredients

- 5 tomatoes
- 6 eggs
- 200 g chicken tenderloin
- 4 to 6 cloves garlic (diced)
- 3 to 4 tbsp stock
- 1 to 2 tsp finely chopped spring onion
- toasted sesames

Seasoning

- 1.5 tsp sugar
- 1 tsp salt

必學不敗竅門

- 煮番茄及炒蛋必須分開烹調，番茄將近煮好時才加入炒蛋，炒蛋才會香滑，番茄也不會太生。
- The tomatoes and eggs must be cooked separately. When the tomatoes are almost done, put in some scrambled eggs. The eggs will still be slightly runny and the tomatoes would not undercook this way.

COOKING TIP

做法

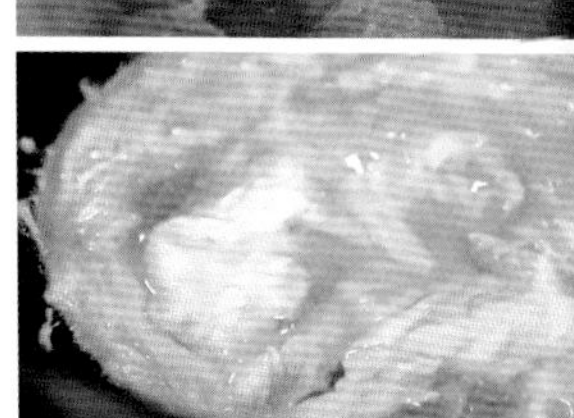

❶ 雞柳洗淨，去筋，剁成蓉，加適量鹽、糖、生粉、生抽及大地魚粉醃勻。用手唧成小肉丸，備用。
❷ 番茄洗淨，剔十字，汆燙後去皮（亦可保留茄皮），切件。
❸ 熱鑊下油，爆香蒜粒，下番茄炒香，加蓋以中大火略煮至變軟及番茄汁釋出，加入調味料拌勻，在鑊邊灒入生抽 1 茶匙，加入上湯，再放入雞肉丸，加蓋以中小火續煮至肉丸熟透。
❹ 同步將雞蛋打入大碗內，撒上少許鹽，待雞肉丸剛煮好，燒熱另一油鑊，倒入蛋，用筷子略攪拌，以中火慢煮至雙色滑蛋。
❺ 將 1/3 滑蛋加入番茄雞肉丸內，餘下滑蛋攪至剛熟，上碟。
❻ 番茄雞肉丸拌勻，放上滑蛋面，撒上葱花、芝麻即成。

Method

❶ Rinse the chicken tenderloin. Trim off the sinews. Finely chop it. Add salt, sugar, potato starch, light soy sauce and ground dried plaice. Mix well. Squeeze out small meatballs in between your thumb and index finger. Set aside.
❷ Rinse the tomatoes. Make a crisscross cut on the bottom. Blanch briefly. Let cool and peel them. (Alternatively, you may skip this step and keep the skin on.) Cut into pieces.
❸ Heat wok and add oil. Stir-fry garlic until fragrant. Add tomatoes and toss until fragrant. Cover the lid and cook over medium-high heat until the tomatoes are tender and juices come out of them. Add seasoning and mix well. Pour in 1 tsp of light soy sauce along the rim of the wok. Add stock and put in the chicken meatballs. Cover the lid. Cook over medium-low heat until the meatballs are cooked through.
❹ Meanwhile, crack the eggs into a big bowl. Sprinkle with a pinch of salt. When the meatballs are cooked, heat another wok and add oil. Pour in the eggs. Stir with a pair of chopsticks gently. Cook over medium heat to make scrambled eggs in two colours.
❺ Transfer 1/3 of the scrambled eggs into the meatball mixture when the eggs are half-set. Then keep on stir-frying the remaining eggs until just cooked through. Transfer onto a serving plate.
❻ Stir the meatballs, tomato and scrambled eggs to mix well. Pour the mixture over the scrambled eggs on a serving plate. Sprinkle with spring onion and sesames on top. Serve.

墨魚 / 鯪魚
cuttlefish/dace

夏日咕嚕魚包肉

Deep-fried dace and cuttlefish balls stuffed with pork in sweet and sour sauce

必學不敗竅門

- 將生果先用熱油浸暖，可代替放入鑊內炒熟的步驟，令生果不易變酸。
- 豬頸肉先剘花、切粒，再釀入魚滑內，可讓肉質保持肉汁、口感爽嫩，更可嘗到魚、豬肉融合後的鮮香。
- It's advisable to pour hot oil on kiwi and peaches instead of stir-frying them in a wok. The fruit is less likely to turn sour this way.
- You can make light crisscross incisions on the pork before dicing it. Then it is stuffed into the minced dace-cuttlefish mixture to keep the pork juicy, springy and tender. You get to taste the complex layering of umami when pork combines with seafood.

COOKING TIP

材料

- 鯪魚滑 200 克
- 豬頸肉 200 克
- 墨魚滑 100 克
- 罐頭開邊桃 2 件
- 奇異果 1 個（綠肉）
- 咕嚕汁 60 毫升
- 士多啤梨果醬 2 茶匙
- 青檸皮半個份量
- 炸芥蘭葉適量（裝飾）

Ingredients

- 200 g minced dace
- 200 g pork collar
- 100 g minced cuttlefish
- 2 canned peach halves
- 1 kiwi
- 60 ml sweet and sour sauce
- 2 tsp strawberry jam
- 1/2 lime (grated zest)
- deep-fried Chinese kale leaves (as garnish)

做法

❶ 奇異果洗淨，去皮切 8 件；罐頭桃切件備用。
❷ 把生果放碗內，淋上熱油先浸暖，用前隔油。
❸ 豬頸肉洗淨，用刀先剞花再切成約 2 厘米方粒，加少許鹽調味。
❹ 墨魚滑與鯪魚滑混合搓勻後撻至起膠。
❺ 取適量魚滑包入豬頸肉，輕揉成肉丸，撲上生粉，備用。
❻ 把肉丸放進熱油炸至微黃，先撈起靜置一會，待熱油餘溫傳至中間肉粒。
❼ 把肉丸放回熱油鑊，炸至金黃，撈起備用。
❽ 熱鑊下適量油，放進咕嚕汁、士多啤梨果醬，加入少許鹽調味，以適量生粉水埋芡，放入肉丸及生果兜炒至「掛汁」，上碟以炸芥蘭葉伴碟，撒上少許青檸皮即成。

Method

❶ Rinse kiwi and peel it. Cut into eighths. Cut the canned peaches into pieces.
❷ Put kiwi and canned peaches into a bowl. Drizzle with hot oil to warm them up. Drain before using.
❸ Rinse the pork. Make light crisscross incisions on the surface. Then cut into 2 cm cubes. Season with a pinch of salt. Mix well.
❹ Mix the minced dace and minced cuttlefish until well incorporated. Lift them off the bowl and slap it back in forcefully. Repeat lifting and slapping until sticky and bouncy in texture.
❺ Wrap a cube of pork into some dace and cuttlefish mixture. Roll into a ball. Coat it in potato starch. Set aside. Repeat this step until all ingredients are used up.
❻ Deep-fry the stuffed dace-cuttlefish balls until lightly browned. Drain and leave them to sit briefly. Allow time for the heat to be conducted to the core.
❼ Put the stuffed dace-cuttlefish balls back in the oil. Deep-fry until golden. Drain.
❽ Heat a wok and add oil. Pour in the sweet and sour sauce, strawberry jam and a pinch of salt. Stir well. Add potato starch thickening glaze and stir until it thickens. Put in the dace-cuttlefish balls, kiwi and peaches. Toss to let the sauce cling on the dace-cuttlefish balls evenly. Transfer onto a serving plate. Garnish with deep-fried kale leaves on the rim of the plate. Sprinkle with grated lime zest. Serve.

豬手
pork trotter

椰青水蓮藕炆豬手

Braised pork trotters with lotus roots and coconut water

必學不敗竅門

- 豬手飛水時加入少量白醋，令豬手燜煮時能保持完整，不容易散爛，而且豬皮更彈牙。
- 蓮藕拍鬆後才炆煮，令蓮藕容易吸收肉汁，會更美味可口。
- When blanching the pork trotters, add a dash of white vinegar. The pork trotters will stay in one piece even after prolonged cooking. The pork skin will be springier in texture after blanched.
- Crushing the lotus roots with the flat side of a knife before cooking them. The lotus roots will taste better that way.

COOKING TIP

材料

- 豬手 600 克
- 蓮藕（泥藕）500 克（約 2 節）
- 紅蘿蔔 1 條
- 椰青 1 個
- 薑 8 片（拍扁）
- 芹菜段適量
- 白醋 2 湯匙

Ingredients

- 600 g pork trotters
- 500 g lotus root (about two segments)
- 1 carrot
- 1 young coconut
- 8 sliced ginger (crushed)
- Chinese celery (cut into shorts lengths)
- 2 tbsp white vinager

調味料

- 甜麵醬 1 湯匙
- 柱侯醬 1.5 湯匙
- 紹興酒 2 湯匙
- 生抽 2 茶匙
- 冰糖碎 2 茶匙
- 蠔油 1 湯匙

Seasoning

- 1 tbsp sweet bean paste
- 1.5 tbsp Chu Hou sauce
- 2 tbsp Shaoxing wine
- 2 tsp light soy sauce
- 2 tsp crushed rock sugar
- 1 tbsp oyster sauce

市面出售的椰青水，買回來後兩小時內必須使用，如不用，必須放於冰格製成冰條，否則會容易變壞。
The fresh young coconut water you get from the market should be used within 2 hours. If you cannot use it in time, freeze it in the freezer. Otherwise, it would go stale very quickly.

如整個椰子買回家，椰子略沖洗後破開，可取得清甜的椰子水。
If you buy a whole coconut home, you can rinse the coconut and cut it open to drain the coconut water right before you use it. That would make sure it is always fresh.

做法

❶ 蓮藕洗淨、去皮，以刀背輕拍鬆，切件。

❷ 將洗淨的豬手放入水內，加入白醋 2 湯匙氽水，盛起，用廚房紙吸乾水分，備用。

❸ 紅蘿蔔洗淨，去皮、切件。

❹ 椰青起肉、切片；椰青水留用。

❺ 熱鑊下油，爆香薑片，放入豬手煎至金黃，加入柱侯醬、甜麵醬、紹興酒、生抽炒勻，再下紅蘿蔔、蓮藕、冰糖，注入熱水約 500 毫升炒勻，轉放入砂煲，下蠔油以中小火炆約 1 小時。

❻ 最後下椰青水再炆 10 分鐘，放進椰青肉、芹菜段，加蓋稍焗一會即可食用。

Method

❶ Rinse lotus root and peel the skin off. Crush gently with the back of a knife. Cut into pieces.

❷ Rinse the pork trotters and put into a pot of water. Add 2 tbsp of white vinegar. Bring to the boil and cook for a while. Drain and set aside the pork trotters. Wipe dry with paper towel.

❸ Rinse and peel the carrot. Cut into pieces.

❹ Open the young coconut. Drain the coconut water and set aside for later use. Scoop out the flesh and slice it. Set aside.

❺ Heat wok and add oil. Stir-fry ginger until fragrant. Put in the pork trotters and fry until golden. Add Chu Hau sauce, sweet bean paste, Shaoxing wine and light soy sauce. Toss well. Add carrot, lotus roots and rock sugar. Pour in about 500 ml of hot water. Toss again. Transfer into a clay pot or casserole. Add oyster sauce. Simmer over medium-low heat for 1 hour.

❻ Add coconut water at last. Simmer for 10 more minutes. Put in coconut flesh and Chinese celery. Cover the lid and turn off the heat. Leave it briefly for flavours to mingle. Serve the whole pot.

大魚頭
head of bighead carp

豬腰瘦肉魚頭煲
Fish head pot with pork kidney and sliced pork

材料

豬腰 1 個
巴西豬脢肉 200 克
大魚頭 1 個
薯粉條 100 克
杜仲黑糯米酒 150 毫升
上湯 200 毫升
薑 6-8 片
指天椒 1 隻
杞子少許（浸軟）
葱段少許

Ingredients

- 1 pork kidney
- 200 g Brazilian pork shoulder butt
- 1 head of bighead carp
- 100 g sweet potato starch noodles
- 150 ml black glutinous rice wine with Du Zhong
- 200 ml stock
- 6 to 8 slices ginger
- 1 bird's eye chilli
- dried goji berries (soaked in water till soft)
- spring onion (cut into short lengths)

做法

1. 溫水浸泡薯粉條約 25-30 分鐘，用前瀝乾水分。
2. 豬腰開邊去筋，表面斜剟十字紋，再斜切薄片，用鹽水浸泡備用。
3. 豬腩肉切片，加適量生抽、糖、生粉及麻油略醃片刻，備用。
4. 大魚頭洗淨，瀝乾水分，開邊，撲上適量生粉。
5. 熱鑊下油，放入薑片爆至微焦黃，將薑盛起放入另一淺湯鍋內。
6. 原鑊放入魚頭，魚臉朝上，煎至金黃即可翻轉另一面煎至金黃，盛起放入湯鍋內，加入上湯。
7. 原鑊放入肉片，快炒至半熟，盛起。
8. 原鑊倒掉油，灒入杜仲黑糯米酒 100 毫升，以餘溫將酒煮暖，再倒入湯鍋內，放入肉片加蓋煮約 5-8 分鐘。
9. 放入薯粉條略煮至軟身，加入適量鹽調味，放入豬腰、指天椒及餘下杜仲黑糯米酒，加蓋煮滚。最後放入杞子及葱段，原鍋享用。

必學不敗竅門

- 想去除豬腰的腥臭味及食時爽脆，首先要將豬腰開邊、去內臟、啤水，並用鹽水浸泡。
- 用酒煮餸時可以分次下酒，第一次下酒可以令食材吸收酒味以達去腥之效；當菜式約八、九成熟時再下酒，可以令整道菜式充滿酒香。
- To remove the unpleasant smell of the pork kidney and to keep it crunchy in texture, cut it in half crosswise. Remove all the connective tissue inside. Then soak it in slowly overflowing water, and soak it in salted water.
- When you make a dish with any wine, you may add the wine in two batches. The first batch is added first and is cooked with other ingredients for prolonged period to let the ingredients pick up the flavours and to remove the gamey taste. When the ingredients are almost done, add wine in second time. Not all alcohol will be cooked off this time and the dish retain a winey aroma.

Method

1. Soak sweet potato starch noodles in warm water for 25 to 30 minutes. Drain right before using.
2. Slice the pork kidney crosswise. Remove any connective tissue and membrane on the cut. Make light crisscross incisions on the smooth side at an angle. Then slice thinly at an angle. Soak in salted water. Drain right before using.
3. Slice the pork. Add light soy sauce, sugar, potato starch and sesame oil. Mix well and leave it briefly.
4. Rinse the fish head. Drain and cut into half crosswise. Coat the fish head in some potato starch.
5. Heat wok and add oil. Stir-fry ginger until lightly browned. Transfer the ginger into another shallow soup pot.
6. In the same wok, put in the fish head with the cut sides down. Fry until golden. Flip to fry the other side until golden. Transfer fish head into the shallow soup pot. Add the stock.
7. In the same wok, stir-fry sliced pork quickly until half-cooked. Set aside.
8. Drain the oil from the wok. Pour in 100 ml of black glutinous rice wine with Du Zhong. Warm the wine with the remaining heat. Pour the wine into the shallow soup pot. Add sliced pork. Cover the lid and cook for 5 to 8 minutes.
9. Put in the sweet potato starch noodles. Cook until they turn soft. Season with salt. Put in pork kidney, bird's eye chilli and the remaining black glutinous rice wine. Cover the lid and bring to the boil. Put in goji berries and spring onion. Serve the whole pot.

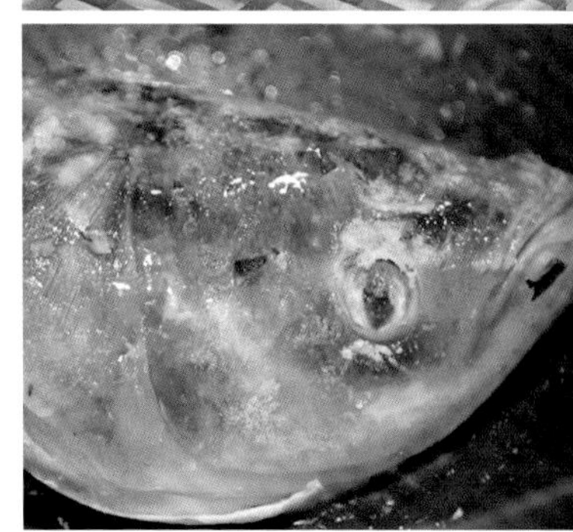

豬肺 / 豬腳筋

pork lung/pork leg sinew

胡椒蘿蔔豬肺豬筋煲

Braised pork lungs and sinews in clay pot with radish

材料

豬肺 1 個
豬腳筋 600 克
白蘿蔔 1 條
青蘿蔔 1/2 條
紅蘿蔔 1/2 條
炸枝竹 2 條
金華火腿上湯 500 毫升
薑 4 片
胡椒粉 1-2 茶匙
大地魚粉 1 茶匙
芹菜段適量

Ingredients

- 1 pork lung
- 600 g pork leg sinews
- 1 radish
- 1/2 green radish
- 1/2 carrot
- 2 deep-fried beancurd sticks
- 500 ml Jinhua ham stock
- 4 slices ginger
- 1 to 2 tsp ground white pepper
- 1 tsp ground dried plaice
- Chinese celery (cut into short lengths)

做法

❶ 三色蘿蔔洗淨、去皮，橫切後再切粗條，蒸軟備用。
❷ 豬肺洗淨，切大件，以白鑊烘乾，期間多次倒掉豬肺流出的水分，直至豬肺完全乾身，盛起備用。
❸ 豬腳筋洗淨後汆水，再蒸腍待用，如較大件可切半。
❹ 炸枝竹先浸軟，再汆水備用。
❺ 將金華火腿上湯倒入砂鍋內加熱，放入薑片、豬腳筋，蓋好煮滾，再放入豬肺及三色蘿蔔，蓋好以中小火再煮 45 分鐘。
❻ 加入炸枝竹，放入大地魚粉及胡椒粉調味，放芹菜段裝飾即成。

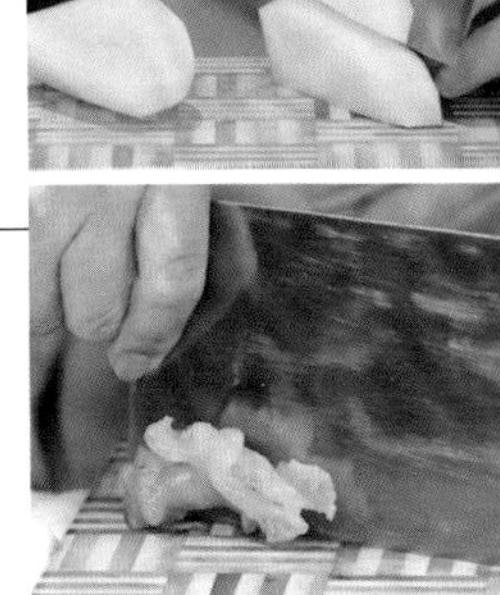

Method

❶ Rinse radish, green radish and carrot. Peel them and cut into segments. Then cut into thick strips. Steam until tender. Set aside.
❷ Rinse the pork lungs. Cut into large chunks. Fry in a dry wok and drain off the liquid in the wok repeatedly until no more liquid oozes from the pork lungs.
❸ Rinse the pork sinews. Blanch in boiling water. Drain and steam until tender. If the sinews are too chunky, cut each in half.
❹ Soak the deep-fried beancurd sticks in water until soft. Blanch in boiling water. Drain.
❺ Pour the Jinhua ham stock into a clay pot. Turn on the heat. Add ginger and pork sinews. Cover the lid and bring to the boil. Put in the pork lungs, radish, green radish and carrot. Cover the lid and simmer over medium-low heat for 45 minutes.
❻ Put in the deep-fried beancurd sticks, ground dried plaice and ground white pepper. Garnish with Chinese celery. Serve the whole pot.

必學不敗竅門

- 蘿蔔順直紋切可以鎖住蘿蔔的水分，亦有助吸收豬骨的味道。
- 豬腳筋可以用牛筋代替。
- 如用鮮枝竹或腐皮代替炸枝竹，注意不要煮太久，否則會溶掉。
- Cutting the radishes and carrot lengthwise helps seal in the moisture. That also helps the radishes and carrot pick up the pork flavour.
- You may use beef sinews instead of pork leg sinews for this recipe.
- If you use fresh beancurd sticks or beancurd sheets instead of deep-fried beancurd sticks, make sure you don't cook them for too long. Otherwise, they may break down and dissolve into the sauce.

OOKING TIPS

豬肉
pork

津白炆豬肉球

Braised pork balls with Napa cabbage

材料

- 津白 1 棵
- 免治豬肉 300 克
- 蝦米 4 湯匙
- 冬菇 5 朵
- 冬菜 2 湯匙
- 薑 2 片
- 果皮碎 1 茶匙
- 上湯 250 毫升
- 杞子 1 茶匙（浸軟）

Ingredients

- 1 head Napa cabbage
- 300 g ground pork
- 4 tbsp dried shrimps
- 5 dried shiitake mushrooms
- 2 tbsp Dong Cai (Tianjian salted cabbage)
- 2 slices ginger
- 1 tsp finely chopped dried tangerine peel
- 250 ml stock
- 1 tsp dried goji berries (soaked in water till soft)

必學不敗竅門

- 搓豬肉球時加入適量蝦米水，可增添香味。
- Adding water used to soak the dried shrimps to the meatballs gives them extra flavours.

COOKING TIP

做法

1. 津白洗淨後切粗條，備用。
2. 蝦米洗淨、浸軟，一半原隻、一半切碎備用；蝦米水加少許薑葱煮開，備用。
3. 冬菜重複浸洗數遍，擠乾水分，備用。
4. 冬菇洗淨、浸軟，去蒂、切片（水留用），備用。
5. 大碗內，放入免治豬肉、蝦米、冬菜，加入鹽 1/4 茶匙、糖 1/2 茶匙及生粉 2 茶匙撈勻，用手搓勻，加入蝦米水約 2 湯匙、麻油 1 茶匙及生抽 1 茶匙撈勻，再撻至起膠，如較乾可酌量加入蝦米水，放入果皮碎拌勻，取適量揉搓成肉球，撲上生粉。
6. 燒熱油鑊，放入肉球煎香至 9 成熟。
7. 燒熱另一油鑊，爆香薑片，放入津白炒香，煮至菜變軟可加入上湯、蝦米水、冬菇水、冬菇片和已爆香原隻蝦米，蓋好炆煮一會，再放入肉球後蓋好煮約 3 分鐘，加入少許胡椒粉及杞子，上碟即成。

Method

1. Rinse the Napa cabbage. Cut into thick strips.
2. Rinse the dried shrimps. Soak in water till soft. Drain and save the soaking water. Set aside half of them. Finely chop the rest. Pour the soaking water into a pot. Add a slice of ginger and some spring onion. Bring to the boil. Strain and set aside.
3. Rinse the Dong Cai repeatedly in water. Squeeze dry.
4. Rinse the shiitake mushrooms. Soak in water till soft. Drain and save the soaking water. Cut off the stems and slice them.
5. In a mixing bowl, mix ground pork, finely chopped dried shrimps and Dong Cai together. Add 1/4 tsp of salt, 1/2 tsp of sugar and 2 tsp of potato starch. Knead and stir with your hand until well combined. Add 2 tbsp of water from soaking dried shrimps, 1 tsp of sesame oil and 1 tsp of light soy sauce. Mix well. Lift the mixture off the bowl and slap it back forcefully for a few times until it turns sticky and resilient. If it feels too dry, add a little water from soaking dried shrimps. Stir in dried tangerine peel. Roll the mixture into meatballs. Coat them in potato starch.
6. Heat wok and add oil. Fry the meatballs until almost done.
7. Heat another wok and add oil. Stir-fry ginger until fragrant. Put in Napa cabbage. Toss until fragrant and tender. Add stock, water from soaking dried shrimps, water from soaking shiitake mushrooms, sliced mushrooms, and whole dried shrimps that have been fried in oil. Cover the lid and simmer for a while. Put in the meatballs. Cover the lid and cook for 3 minutes. Sprinkle with ground white pepper and dried goji berries. Serve.

花膠

dried fish maw

章魚花膠原個節瓜煲

Braised whole Chinese marrows with octopus and fish maw in clay pot

必學不敗竅門

- 用作浸發花膠的器皿一定要清洗乾淨，沒有油分，花膠才能浸發成功。
- 無論是炮製節瓜煲或節瓜湯，記得要整個節瓜放入煲內烹調，享用時才將節瓜切開；因為瓜肉已吸收其他材料的精華，所以特別鮮甜，口感軟糯。
- To rehydrate dried fish maw, make sure the container is perfectly clean without the tiniest bit of grease. Otherwise, the fish maw will dissolve in the water.
- When you make soup or any dish with Chinese marrow, it's advisable to cook it in whole. Slice it right before serving. You can keep the tender texture of the Chinese marrow this way, while letting its flesh pick up the flavours of other ingredients.

COOKING TIP

材料

- 挪威花膠（約 17 頭）1 隻
- 節瓜 2 個
- 章魚 1 隻
- 蠔豉 8 粒
- 花生 60 克
- 眉豆 60 克
- 豬骨 500 克
- 冬菇 8 朵
- 紅棗 5 粒
- 上湯 1 公升

Ingredients

- 1 Norwegian dried fish maw (17-headed)
- 2 Chinese marrows
- 1 dried octopus
- 8 dried oysters
- 60 g peanuts
- 60 g black-eyed beans
- 500 g pork bones
- 8 dried shiitake mushrooms
- 5 red dates
- 1 litre stock

做法

1. 花生及眉豆洗淨、浸軟，備用。
2. 花膠浸發步驟：先乾蒸 10 分鐘，再用凍水充分浸軟，以薑、葱、酒加開水浸泡過夜；發好切 5-6 件，待用。
3. 豬骨以 1/2 湯匙粗鹽先醃過夜，用前先汆水。
4. 節瓜洗淨，用刀背刮去深綠色外皮，原個備用。
5. 冬菇浸軟（水留用），去蒂備用；紅棗洗淨、去核，備用。
6. 章魚及蠔豉先浸軟，吸乾水後煎香；章魚切成 5-6 件，備用。
7. 上湯放鍋內，加入花生、眉豆、豬骨、節瓜、冬菇、紅棗，用中大火煮滾，加入章魚及蠔豉，蓋好煲 1 小時。
8. 開蓋加入冬菇水及花膠，再煲 30 分鐘即成。

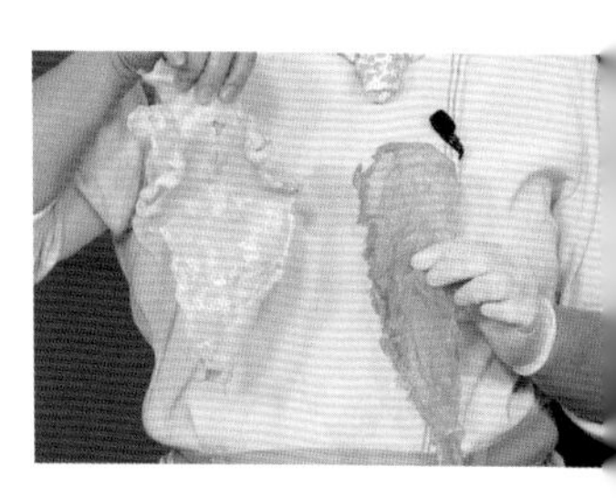

花膠浸發前（右）；浸發後（左）。
Fish maw before (R) and after (L) rehydration.

Method

1. Rinse peanuts and black-eyed beans. Soak them in water till soft. Drain.
2. To rehydrate the fish maw, steam the dry fish maw for 10 minutes. Soak it in cold water until soft. Then soak it in drinking water with ginger, spring onion and wine overnight. Drain. Cut the fish maw into 5 to 6 pieces.
3. Rub 1/2 tbsp of coarse salt over the pork bone evenly. Leave it overnight in the fridge. Blanch in boiling water and drain before using.
4. Rinse the Chinese marrows. Scrape off the dark green skin with the back of a knife. Keep them in whole.
5. Soak shiitake mushrooms in water till soft. Drain and set aside the soaking water. Cut off the stems. Set aside. Rinse the red dates and stone them.
6. Soak dried octopus and dried oysters in water till soft. Drain and wipe dry. Fry them in a little oil until fragrant. Cut the octopus into 5 to 6 pieces.
7. Pour stock into a clay pot. Add peanuts, black-eyed beans, pork bones, Chinese marrows, shiitake mushrooms and red dates. Bring to the boil over medium-high heat. Add dried octopus and dried oysters. Cover the lid and boil for 1 hour.
8. Add water from soaking shiitake mushrooms and fish maw. Cover the lid and cook for 30 minutes. Serve the whole pot.

雞油金腿煎焗薯仔片

Scalloped potato with Jinhua ham fried in chicken fat

材料

- 黃肉馬鈴薯 2 個
- 金華火腿 120 克
- 鮮雞油 3 湯匙
- 薑 2 片（拍扁）
- 上湯 100 毫升
- 葱花少許

Ingredients

- 2 yellow-fleshed potatoes
- 120 g Jinhua ham
- 3 tbsp freshly rendered chicken fat
- 2 slices ginger (crushed)
- 100 ml stock
- finely chopped spring onion

做法

1. 馬鈴薯去皮、切片，用淡鹽水浸泡備用。
2. 金華火腿汆水後切片，瀝乾水分備用。
3. 用雞皮或臀部，慢火乾煎至出雞油，備用。
4. 燒熱鑊，下鮮雞油，先爆香薑片，下金華火腿片爆炒，下薯片大火快炒，逐少加入上湯炒勻，將火腿片墊底，蓋上煮 8 分鐘至薯片熟透微焦香，加入糖少許調味，兜勻上碟，最後撒上葱花即成。

Method

1. Peel and slice potatoes. Soak them in lightly salted water. Drain before using.
2. Blanch Jinhua ham in boiling water. Slice it. Drain well.
3. Render chicken fat from chicken skin or chicken tail. Fry over low heat in a dry wok until oil is rendered. Drain and set aside the chicken fat.
4. Heat a wok. Pour in the chicken fat. Stir-fry ginger until fragrant. Put in Jinhua ham and toss well. Put in the sliced potatoes. Toss quickly over high heat. Slowly stir in the stock while tossing. Push the sliced Jinhua ham to line the wok. Cover the lid and cook for 8 minutes until the potatoes are cooked through and lightly browned. Season with a pinch of sugar. Toss and save on a serving plate. Garnish with spring onion. Serve.

必學不敗竅門

- 煎焗時，金華火腿要墊底，才可以煮出火腿甘香味道，令薯片更美味。
- Before you cover the lid and cook the potatoes through, make sure you push to line the wok with Jinhua ham. That brings out the robust meaty taste of the ham and make the potatoes taste even better.

COOKING TIPS

莧菜
amaranth

翡翠苗餃子

Pork and amaranth dumplings

材料

莧菜 120 克
免治豬肉 200 克
餃子皮 20-30 張
榨菜 40 克
馬蹄 5 粒
蝦米 1 湯匙

Ingredients

- 120 g amaranth
- 200 g ground pork
- 20 to 30 sheets dumpling skin
- 40 g Zha Cai (Sichuan-style preserved mustard tuber)
- 5 water chestnuts
- 1 tbsp dried shrimps

餃子餡調味料

大地魚粉 1 湯匙
胡椒粉 1/2 茶匙
生粉 1 茶匙
糖 1/2 茶匙
生抽 1/2 湯匙
麻油 1 湯匙

Seasoning for filling

- 1 tbsp ground dried plaice
- 1/2 tsp ground white pepper
- 1 tsp potato starch
- 1/2 tsp sugar
- 1/2 tbsp light soy sauce
- 1 tbsp sesame oil

做法

❶ 莧菜洗淨、去根，取菜梗切粒，備用。榨菜洗淨、索乾水分，切粒備用。

❷ 馬蹄洗淨、去皮，切碎備用。蝦米洗淨、浸軟，切碎備用。

❸ 大碗內，放入免治豬肉、榨菜粒、馬蹄碎、蝦米碎，下大地魚粉、胡椒粉、生粉、糖、生抽、麻油撈勻，再輕撻至起膠，加入莧菜粒充分拌勻成餡料，備用。

❹ 取適量餡料，放入餃子皮內，包好後以水封口，按個人喜好烹煮。

❺ 煎餃：燒熱平底鍋，油燒熱後放入餃子，煎至底部金黃，倒入適量上湯（浸過餃子底部即可），蓋好鍋蓋，煎煮至水分蒸發，即可上碟。

❻ 水煮餃：沸水內加入餃子，待水沸騰，加入約半份凍水，再煮至沸騰即可，食時可配以上湯享用。

Method

❶ Rinse amaranth. Cut off the roots. Pluck off the leaves and use the stems only. Dice the stems. Set aside. Rinse Zha Cai well. Wipe dry and dice it.

❷ Rinse water chestnuts. Peel and chop finely. Set aside. Rinse and soak dried shrimps in water till soft. Finely chop them.

❸ In a mixing bowl, put in ground pork, Zha Cai, water chestnuts and dried shrimps. Add seasoning for filling. Mix well. Lift the mixture off the bowl and slap it back forcefully a few times until sticky and resilient. Add diced amaranth stems. Mix well. This is the filling.

❹ Wrap some filling in a sheet of dumpling skin. Fold and seal the seam with water. You may then make fried dumplings or dumpling soup as you prefer.

❺ For fried dumplings, heat a pan and add oil. Put in the dumplings and fry until the bottoms are golden. Pour in some stock (just enough to cover the base of the pan). Cover the lid. Cook until the stock is cooked off. Serve.

❻ For dumpling soup, boil water in a pot. Put in the dumplings and cook until the water boils again. Add half as much cold water to the pot. Bring to the boil again and the dumplings are done. Drain and serve the dumplings in stock.

必學不敗竅門

- 水煮餃子滾起時加入凍開水，可保證餡熟而餃子皮不會穿破。
- To cook the dumplings in water, add cold boiled water after it boils. This would ensure the dumplings are cooked through and their skin won't break.

OOKING TIPS

港式麻婆豆腐

Ma Po tofu in Hong Kong Style

材料

- 盒裝蒸煮豆腐 1 盒
- 免治牛肉 200 克
- 榨菜 40 克
- 海參 1/2 條（已浸發）
- 蒜粒 1 湯匙
- 花椒油 2 茶匙
- 豆瓣醬 3 茶匙
- 辣椒油 2 茶匙
- 葱花少許（裝飾）
- 紅椒絲少許（裝飾）

Ingredients

- 1 pack soft tofu for steaming
- 200 g ground beef
- 40 g Zha Cai (Sichuan-style preserved mustard tuber)
- 1/2 sea cucumber (rehydrated)
- 1 tbsp diced garlic
- 2 tsp Sichuan peppercorn oil
- 3 tsp spicy bean sauce
- 2 tsp chilli oil
- finely chopped spring onion (as garnish)
- shredded red chillies (as garnish)

做法

❶ 免治牛肉加入油 2 茶匙、生粉 1 茶匙、生抽 1 湯匙及糖少許撈勻，略醃，用前加入少許水拌開。
❷ 豆腐切成方粒，放入熱水內（水蓋豆腐面即可），加入鹽少許及老抽 3-5 湯匙令豆腐上色，備用。
❸ 榨菜洗淨，浸泡、沖洗幾遍，切粒；海參切粒，備用。
❹ 燒熱油鑊，先放入蒜粒及榨菜爆香，再加入海參粒炒勻，灒少許紹興酒，加入免治牛肉，下鹽少許、糖半茶匙、豆瓣醬、花椒油及辣椒油調味，轉小火兜勻，將豆腐隔水加入內，輕輕拌勻，綴以紅椒絲，上碟，撒上葱花即成。

Method

❶ Add 2 tsp of oil, 1 tsp of potato starch, 1 tbsp of light soy sauce and a pinch of sugar to the ground beef. Mix well and leave it briefly. Stir in a little water right before using.
❷ Dice the tofu. Soak in hot water (enough water to cover). Add a pinch of salt and 3 to 5 tbsp of dark soy sauce to colour the tofu.
❸ Rinse Zha Cai. Soak in water. Rinse repeatedly. Then dice it and set aside. Dice the sea cucumber.
❹ Heat wok and add oil. Stir-fry garlic and Zha Cai until fragrant. Put in sea cucumber and toss well. Drizzle with Shaoxing wine. Add ground beef, a pinch of salt, 1/2 tsp of sugar, spicy bean sauce, Sichuan peppercorn oil and chilli oil. Turn to low heat and toss well. Drain the tofu and add to the wok. Stir gently to mix well. Garnish with red chillies. Transfer onto a serving plate. Garnish with spring onion. Serve.

必學不敗竅門

- 將豆腐粒放入熱水浸泡及加老抽上色，可令豆腐定形，烹調時不易散開。
- It is advisable to soak the diced tofu into hot water and add dark soy sauce to colour it. This step would firm up the tofu and it won't break down as easily when cooked with other ingredients.

COOKING TIPS

海鮮
seafood

梅子海鮮波子飯

Plum-scented assorted seafood with deep-fried rice balls

材料

- 法國白鱈魚 1 片（約 400 克）
- 蝦 6 隻
- 蜆 10 隻
- 梅子 3-4 粒
- 青檸 1 個
- 白飯 1 碗
- 炒香芝麻適量
- 上湯 500 毫升
- 香茅 1 條
- 薑 3 片

Ingredients

- 1 French cod fillet (about 400 g)
- 6 shrimps
- 10 live clams
- 3 to 4 dried plums (stoned)
- 1 lime
- 1 bowl steamed rice
- toasted sesames
- 500 ml stock
- 1 stem lemongrass
- 3 slices ginger

調味料

- 大地魚粉 1/2 茶匙
- 魚露 1/2 茶匙

Seasoning

- 1/2 tsp ground dried plaice
- 1/2 tsp fish sauce

做法

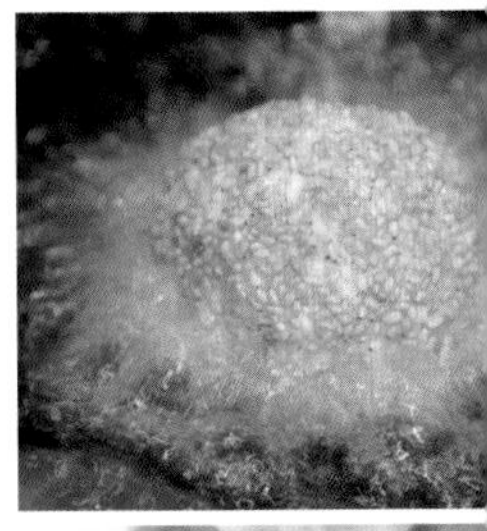

❶ 蜆用海水浸泡，放雪櫃冷藏約 4 小時，吐沙後洗淨，焯至半熟及殼張開，盛起，瀝乾水分。
❷ 蝦洗淨，瀝乾水分，去殼、去頭及去腸，備用。
❸ 白鱈魚去皮、切粒，加少許胡椒粉、鹽及生粉略醃，備用。
❹ 梅子切碎；香茅拍扁、切段；青檸半個榨汁，半個切片，備用。
❺ 白飯用保鮮紙包裹，搓成球狀，均勻地蘸上芝麻，下油鑊炸至金黃色，上碟。
❻ 熱鑊下油，爆香薑片，放入白鱈魚粒煎至微金黃，盛起。
❼ 原鑊加入上湯、梅子碎、香茅段、青檸汁，加糖 1 茶匙調味煮滾，下蜆、青檸片，加入調味料。待蜆熟後，加入蝦、鱈魚粒，煮滾後上碟即成。

Method

❶ Soak clams in salted water. Refrigerate for 4 hours for them to spit out the sand. Rinse well and blanch in boiling water until they open and half cooked. Drain.
❷ Rinse the shrimps. Drain. Shell, devein and remove the heads.
❸ Skin the cod fillet. Dice it. Sprinkle with a pinch of salt, ground white pepper and potato starch. Mix well and leave it briefly.
❹ Finely chop the dried plums. Bruise the lemongrass with the back of a knife. Cut into short lengths and set aside. Cut the lime in half. Squeeze half of it. Slice the rest.
❺ Wrap the rice in cling film. Roll into a ball. Coat the rice ball in white sesames evenly. Deep-fry in oil until golden. Drain and save on a serving plate.
❻ Heat wok and add oil. Stir-fry ginger until fragrant. Fry the diced cod until lightly browned on all sides. Set aside.
❼ In the same wok, pour in stock. Add dried plums, lemongrass, lime juice and 1 tsp of sugar. Bring to the boil. Put in clams, sliced lime and seasoning. When the clams are cooked through, put in shrimps and diced cod. Bring to the boil and serve.

必學不敗竅門

- 製作波子飯並不難，用保鮮紙包着搓成飯糰，又圓又靚！
- It's not hard to make rice balls at all. Just wrap some rice in cling film. Twist and roll into balls. They are always perfectly round and beautiful.

OOKING TIPS

蓮藕

lotus root

蓮漪飄香

Deep-fried lotus root sandwich with cuttlefish and cheese filling

做法

1. 處理桂花梨：雪梨去皮、去芯，切半後放入煲內，加入紅麴米、桂花糖及適量水（蓋過梨面），小火煮 1 小時，完成後冷藏，切粒備用。
2. 蓮藕洗淨，去皮，切薄片，備用。
3. 芝士 2 片疊起，每片切成 16 小方片，備用。
4. 取蓮藕片撲上生粉，均勻地塗上墨魚膠（需填滿蓮藕孔），放入一小片芝士，再塗上墨魚膠，取另一片蓮藕撲上生粉，蓋上夾好，邊位用墨魚膠黏好，再撲上生粉，蘸上炸漿。
5. 燒熱油鑊，放入蓮藕夾炸至金黃香脆，盛起。
6. 取小鍋燒熱少許油，加入士多啤梨汁所有材料拌勻，煮熱即可。
7. 碟內放上桂花梨，再加入蓮藕夾，淋上士多啤梨汁即成。

必學不敗竅門

- 蓮藕先蘸上生粉才釀入墨魚膠，並用墨魚膠黏好邊位，炸時不容易散開。
- You have to coat the lotus root in potato starch before stuffing it with minced cuttlefish. Make sure the filling goes all the way to the edges to bond the lotus root well. That would ensure the lotus-root sandwiches won't break apart when deep-fried.

COOKING TIP

材料

- 蓮藕 1 節
- 墨魚膠 200 克
- 片裝芝士 2 片
- 炸漿 150 毫升

Ingredients

- 1 segment lotus root
- 200 g minced cuttlefish
- 2 slices cheddar cheese
- 150 ml deep-frying batter

士多啤梨汁

- 士多啤梨果醬 4 湯匙
- 糖醋汁 3 湯匙
- 檸檬汁 2 湯匙

Strawberry glaze

- 4 tbsp strawberry jam
- 3 tbsp sweet and sour sauce
- 2 tbsp lemon juice

桂花梨

- 雪梨 1 個
- 紅麴米 1 湯匙
- 桂花糖 100 毫升

Osmanthus poached pear

- 1 Chinese pear
- 1 tbsp red yeast rice
- 100 ml candied osmanthus

Method

1. To make the osmanthus poached pear, peel and core the pear first. Cut in half and put into a small pot. Add red yeast rice and candied osmanthus. Pour in enough water to cover. Turn on low heat and simmer for 1 hour. Leave it to cool and refrigerate. Dice it.
2. Rinse the lotus root. Peel and slice thinly across the length.
3. Stack the 2 slices of cheese. Cut each into 16 small squares.
4. Coat a slice of lotus root in potato starch. Smear minced cuttlefish evenly on it. Make sure all holes are covered. Put on a small piece of cheese. Spread another layer of minced cuttlefish over. Coat another slice of lotus root in potato starch and stack it over the minced cuttlefish filling. Press gently and make sure the filling goes all the way to the rim. Smear the filling along the rim to ensure good bonding. Coat the sandwich in potato starch again. Dip into the deep-frying batter. Repeat this step until all ingredients are used up.
5. Heat wok and add oil. Deep-fry the lotus root sandwich until golden and crispy. Drain and set aside.
6. Heat a small pot and add a little oil. Pour in the strawberry glaze ingredients and mix well. Bring to the boil. Set aside.
7. Arrange the osmanthus poached pear on a serving plate. Put the deep-fried lotus root sandwiches on top. Drizzle with strawberry glaze. Serve.

蟹肉
crabmeat

蝦蟹豆腐卷

Steamed beancurd skin rolls with tofu and crabmeat filling

材料

- 即食蟹肉 80 克
- 盒裝蒸煮豆腐 1 盒
- 腐皮 1 張（圓形）
- 鮑魚汁 4 湯匙
- 麻油 1 茶匙
- 即食蟹子 4 茶匙
- 炒熟蝦子 2 茶匙

Ingredients

- 80 g instant crabmeat
- 1 pack soft tofu for steaming
- 1 sheet beancurd skin (round)
- 4 tbsp abalone sauce
- 1 tsp sesame oil
- 4 tsp tobikko (flying fish roe)
- 2 tsp toasted shrimp roe

做法

❶ 豆腐橫切半後再切成 8 件，備用。
❷ 腐皮對摺剪開，重複至裁成 8 等分，放入熱油鑊略炸軟，盛起，浸入凍開水內（水要浸過面），待一會至腐皮吸透水分變軟，備用。
❸ 小心取出一片腐皮，放毛巾上吸乾水分，排上一件豆腐、適量蟹肉放面，豆腐表面沾少許鹽，捲成筒狀，放上蒸碟以大火蒸 7-8 分鐘。
❹ 另燒熱少許油，放入鮑魚汁、麻油煮滾，下適量生粉芡，趁熱淋上豆腐卷，放上蟹子及蝦子即成。

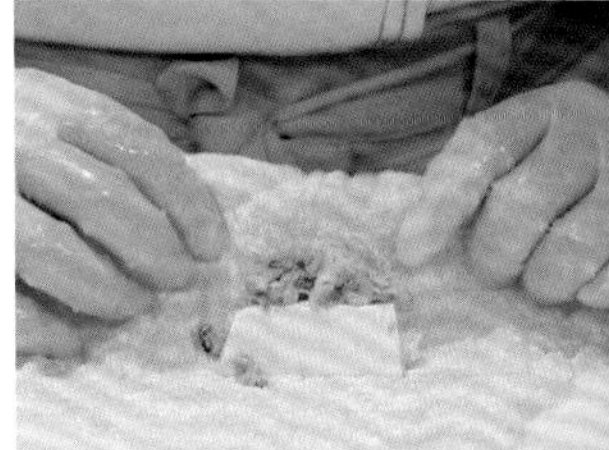

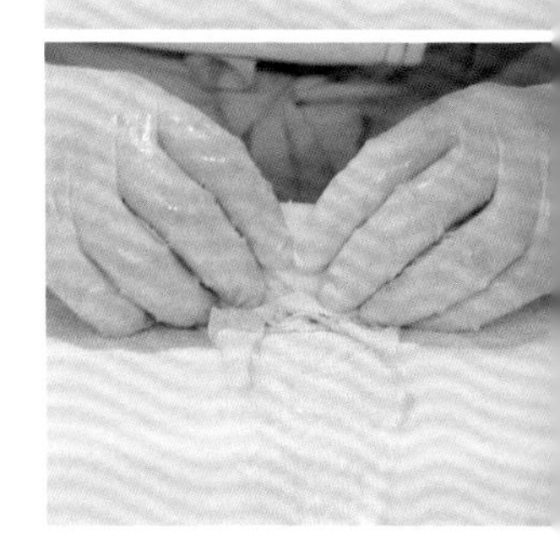

Method

❶ Slice the tofu in half crosswise. Then cut into 8 pieces.
❷ Fold the beancurd skin in half. Cut it. Fold in half and cut again. Repeat until you get 8 equal pieces. Deep-fry the beancurd skin in oil until softened. Drain. Soak in cold drinking water (enough to cover). Wait till the beancurd skin picks up the water and turn soft and foldable. Drain.
❸ Carefully lift a piece of beancurd skin. Lay it flat on a clean towel to absorb water. Arrange a piece of tofu and some crabmeat on top. Sprinkle a pinch of salt over the tofu. Then fold the beancurd skin and roll into a spring roll. Arrange on a steaming plate. Steam over high heat for 7 to 8 minutes.
❹ In another wok, heat some oil. Pour in the abalone sauce and sesame oil. Bring to the boil. Stir in potato starch thickening glaze. Pour the glaze over the steamed beancurd skin rolls. Arrange tobikko and shrimp roe on top. Serve.

必學不敗竅門

- 腐皮先炸脆，再放入凍開水浸透，捲時不易散開，而且味道更香。
- Deep-fry the beancurd skin first and then soak it in cold drinking water. That would make the beancurd skin more flexible and the rolls are less likely to go loose. It also tastes better.

OOKING TIPS

牛柳
beef tenderloin

乾炒牛河

Beef "Chow Fun" (Cantonese fried rice noodles with beef)

做法

示範短片

1. 牛柳切薄片，加入醃料拌勻，略醃備用。
2. 燒熱油鑊，放入牛柳大火輕輕煎香，見稍轉色、半熟即盛起。
3. 燒熱油鑊，放入芽菜大火略煸炒，盛起備用。
4. 用筷子把河粉撥散；燒熱油鑊（油稍多點），放入河粉輕煎，不要翻動，灑入少許鹽，轉動鑊令河粉受熱均勻，煎透後反轉另一面，輕翻炒至軟身。
5. 灒入鮮醬油及生抽，再加入韭黃快炒，將牛柳及芽菜回鑊快炒，加入葱段及老抽拌勻，上碟後放上蛋絲及紅椒絲，灑上芝麻即成。

必學不敗竅門

- 炒河粉時，緊記鑊必須熱透，下河粉後不要亂動，先煎香一面至金黃色才翻轉繼續煎香另一面，以免弄得散碎。
- When you stir-fry rice noodles, make sure the wok must be hot enough. Do not stir the noodles after putting them in. Fry until one side is golden before flipping to fry the other side. Stirring and tossing too much would break the noodles into bits.

COOKING TIP

材料	Ingredients
牛柳 250 克	250 g beef tenderloin
河粉（炒）500 克	500 g ribbon rice noodles for stir-frying
芽菜 100 克	100 g mung bean sprouts
韭黃 40 克（切段）	40 g yellow chives (cut into short lengths)
葱段 2 條	2 spring onion
蛋絲少許	shredded egg omelette
紅椒絲少許	red chillies (shredded)
炒香芝麻少許	toasted sesames

醃料	Marinade
油 3 茶匙	3 tsp oil
生粉 1/2 茶匙	1/2 tsp potato starch
糖少許	sugar
薑汁 2 湯匙	2 tbsp ginger juice
紹興酒 1 茶匙	1 tsp Shaoxing wine

調味料	Seasoning
生抽 1 湯匙	1 tbsp light soy sauce
老抽 1 湯匙（用等份水拌勻）	1 tbsp dark soy sauce (mixed with equal volume of water)
鮮醬油 2 茶匙	2 tsp Maggi seasoning

Method

1. Slice the beef thinly. Add marinade and mix well. Leave it briefly.
2. Heat wok and add oil. Put in the beef and stir-fry over high heat until half-cooked and it changes colour slightly. Set aside.
3. Heat wok and add some oil, sauté mung bean sprouts slightly. Set aside.
4. Separate the rice noodles with a pair of chopsticks. Heat wok and add a bit more oil than usual. Put in the rice noodles and fry them. Do not stir them. Sprinkle with a pinch of salt. Swirl the wok to heat the noodles evenly. Flip the noodles after one side is heated through. Gently toss them until soft.
5. Drizzle with Maggi seasoning and light soy sauce. Add yellow chives and toss quickly. Put the beef and mung bean sprouts in. Toss quickly. Add spring onion and dark soy sauce. Mix well. Transfer onto a serving plate. Arrange shredded egg omelette on top and garnish with red chillies. Sprinkle with sesames. Serve.

牛仔骨

beef short rib

中式牛仔骨

Chinese-style beef short ribs

材料

- 蘇格蘭牛仔骨 3 件
- 啤酒 350 毫升
- 洋葱 1 個（切絲）
- 番茄 1 個（切細粒）
- 白蘭地 1 茶匙

Ingredients

- 3 pieces Scottish beef short ribs
- 350 ml beer
- 1 onion (shredded)
- 1 tomato (diced)
- 1 tsp brandy

醬汁

- 喼汁 5 湯匙
- 薑汁 2 湯匙
- 鮮醬油 1 湯匙
- 片糖碎 1.5 湯匙
- 咕嚕汁 1 湯匙
- 生粉 1/2 茶匙

Sauce

- 5 tbsp Worcestershire sauce
- 2 tbsp ginger juice
- 1 tbsp Maggi seasoning
- 1.5 tbsp crush raw cane sugar slab
- 1 tbsp sweet and sour sauce
- 1/2 tsp potato starch

做法

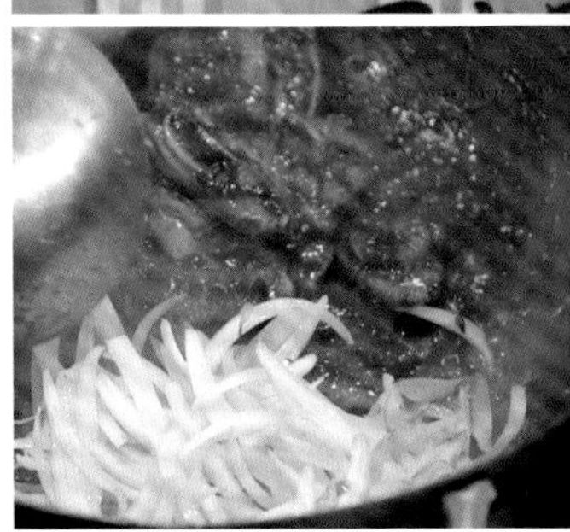

❶ 牛仔骨解凍，洗淨，吸乾水分，每件切成 3 大片，浸泡啤酒半小時，瀝乾，加入鹽少許及生粉適量拌勻，備用。
❷ 所有醬汁材料拌勻，備用。
❸ 燒熱油鑊，放入洋葱爆炒至微焦，盛起備用。
❹ 原鑊放入牛仔骨，大火煎香至兩面微黃。
❺ 同一時間，燒熱另一油鑊，放入番茄粒炒香，加入醬汁後轉小火煮滾，放入牛仔骨，轉大火炒勻收汁，放入洋葱，灒入白蘭地，上碟即成。

Method

❶ Thaw the beef short ribs. Rinse and wipe dry. Cut each piece into three large slices. Soak in beer for 30 minutes. Drain. Sprinkle with a pinch of salt and potato starch. Mix well.
❷ Mix all sauce ingredients together to combine.
❸ Heat wok and add oil. Stir-fry onion over high heat until lightly browned. Set aside.
❹ In the same wok, put in the beef short ribs. Fry over high heat until both sides lightly browned.
❺ Meanwhile, heat oil in another wok. Stir-fry tomato until fragrant. Add the sauce ingredients and turn to low heat. Bring to the boil. Put in the beef and turn to high heat. Toss until the sauce reduces. Add onion and drizzle with brandy. Serve.

必學不敗竅門

- 牛仔骨煮前放入啤酒浸約半小時，肉質更鬆軟。
- Soaking the beef in beer for 30 minutes helps tenderize the meat fibres. The beef will turn out more tender.

OOKING TIPS

帶子 / 雞蛋
scallop / egg

夜香花帶子炒滑蛋

Creamy scrambled eggs with scallops and night willow herb

必學不敗竅門

- 帶子用蝦眼水略焯及焗片刻，印乾水分，可固定形狀，以及保持帶子中心肉質嫩滑。
- To par-boil the scallops, heat water until tiny bubbles start to appear. Put in the scallops and cover the lid briefly. Drain and wipe dry. This step would firm up the scallops slightly to retain their shapes. The core of the scallops remains juicy and tender.

夜香花有清肝明目的作用，常用於煲冬瓜盅及煎雞蛋。
Night willow herb is said to clear the Liver meridians and improve eyesight. It is commonly used in braised soup in whole winter melon and omelettes.

COOKING TIPS

材料

- 夜香花 80 克
- 帶子 7 粒
- 雞蛋 6 隻
- 生粉適量
- 紅椒粒少許
- 炒香芝麻少許

Ingredients

- 80 g floral buds of night willow herb
- 7 scallops
- 6 eggs
- potato starch
- red chillies (diced)
- toasted sesames

做法

1. 帶了解凍、洗淨，以攝氏 70 度熱水浸 1 分鐘至定形，盛起，以廚房紙吸乾水分，撲上生粉，備用。
2. 熱鑊下油，用大火放入帶子煎至兩面金黃，盛起備用。
3. 夜香花洗淨，瀝乾水分備用。
4. 雞蛋打進大碗內（不用拂勻），撒上適量幼鹽。
5. 熱鑊下油，倒入蛋液，一邊以筷子略攪拌蛋液，一邊加入夜香花、帶子煮成滑蛋，上碟後撒上芝麻，並以紅椒粒伴碟即成。

Method

1. Thaw the scallops and rinse well. Soak them in hot water at 70°C for 1 minute to firm them up slightly and retain their shapes. Drain and set aside. Wipe dry with paper towel. Coat them in potato starch. Set aside.
2. Heat wok and add oil. Sear the scallops over high heat until both sides golden. Set aside.
3. Rinse the night willow herb. Drain.
4. Crack the eggs into a mixing bowl. Do not whisk them. Sprinkle with a pinch of table salt.
5. Heat wok and add oil. Pour in the eggs. Stir with eggs briefly with chopsticks while adding the night willow herb and scallops. When the eggs are half set, transfer onto a serving plate. Sprinkle with sesames and garnish with red chillies. Serve.

花蛤

Venus clam

私房梅酒煮花蛤

Steamed clams in plum wine sauce

材料

- 花蛤 600 克
- 梅酒 200 毫升
- 梅子肉（浸過梅酒）5 粒
- 芹菜 1 棵（切粒）
- 芫茜 2 棵（切碎）
- 蒜粒 3 湯匙
- 上湯 200 毫升
- 玫瑰露 1 湯匙
- 黑椒碎 1/2 茶匙
- 胡椒粉 1/2 茶匙
- 紅椒絲少許

Ingredients

- 600 g Venus clams
- 200 ml plum wine
- 5 plums (soaked in wine previously)
- 1 sprig Chinese celery (diced)
- 2 sprigs coriander (finely chopped)
- 3 tbsp diced garlic
- 200 ml stock
- 1 tbsp Chinese rose wine
- 1/2 tsp ground black pepper
- 1/2 tsp ground white pepper
- red chillies (shredded)

做法

1. 梅子肉切碎，備用。
2. 花蛤用海水浸泡，放雪櫃冷藏約 4 小時，吐沙後洗淨（丟掉沒張開殼的花蛤），放入沸水焯至殼打開，半熟即可盛起，瀝乾水分。
3. 熱鑊下油，爆香蒜粒，下梅子肉、上湯，滾起後下半份量梅酒及花蛤，加黑椒碎調味，滾起後下剩餘的梅酒、芹菜粒、芫茜碎，最後灒入玫瑰露，加入胡椒粉及以紅椒絲裝飾，上碟即成（可放入原粒梅子墊在碟底）。

Method

1. Stone the plums and finely chop them.
2. Soak the clams in salted water. Refrigerate for 4 hours for them to spit out the sand. Rinse and blanch in boiling water until they open slightly and are half-cooked. Discard any clam that does not open. Set aside and drain well.
3. Heat wok and add oil. Stir-fry garlic until fragrant. Put in the plums and stock. Bring to the boil. Put in half of the plum wine and all clams. Season with ground black pepper. Bring to the boil again and add the remaining plum wine, Chinese celery and coriander. Drizzle with Chinese rose wine and sprinkle with ground white pepper. Garnish with red chillies. Save on a serving plate and serve. (Optionally, put a few whole plums on the bottom of the plate and pour the clams over.)

必學不敗竅門

- 煮花蛤前放入大滾水微焯至殼半開，即撈起，確保乾淨及全部新鮮，死掉的即丟掉。
- Before cooking the clams, blanch them in vigorously boiling water until they open slightly. Then drain and set aside. This step would make sure all clams are clean and fresh. Discard those dead ones so that they won't contaminate others.

軟殼蟹
soft-shelled crab

綠豆咖喱軟殼蟹
Soft-shelled crab curry with mung beans

材料

- 軟殼蟹 2 隻
- 開邊綠豆 60 克
- 鮮菠蘿 100 克（切粒）
- 鮮菠蘿 2 湯匙（切幼粒）
- 洋葱 1/2 個（切粒）
- 番茄 2 個（切粒）
- 芫茜 1 棵

醬料

- 咖喱醬 2 湯匙
- 花生醬（無粒）1 茶匙
- 鮮椰漿 120 毫升
- 花奶 60 毫升
- 雞蛋 1 隻

Ingredients

- 2 soft-shelled crabs
- 60 g split mung beans
- 100 g fresh pineapple (peeled and diced)
- 2 tbsp fresh pineapple (peeled and diced finely)
- 1/2 onion (diced)
- 2 tomatoes (diced)
- 1 sprig coriander

Sauce

- 2 tbsp curry paste
- 1 tsp peanut butter (creamy)
- 120 ml freshly squeezed coconut milk
- 60 ml evaporated milk
- 1 egg

做法

1. 軟殼蟹解凍、洗淨，去鰓、去厴，用廚房紙印乾水分，每隻斬成兩件，加入蛋白半隻及適量生粉拌勻備用。
2. 開邊綠豆洗淨、浸軟，隔水蒸半小時至腍，備用。
3. 熱鑊下油，下軟殼蟹炸至熟透，瀝乾油分，上碟。
4. 熱鑊下油，下洋葱爆香，加入番茄粒翻炒，下菠蘿幼粒、咖喱醬爆香，再加入開邊綠豆、菠蘿粒、花生醬，細火慢煮至濃稠，最後加入鮮椰漿及花奶，下蛋液拌勻，淋在軟殼蟹上，以芫茜裝飾即成。

Method

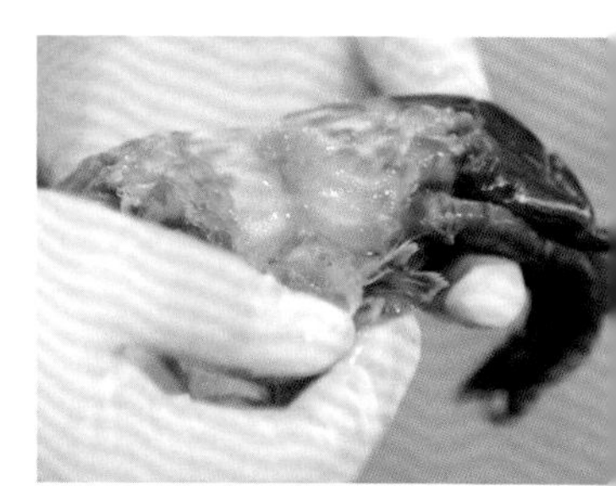

1. Thaw the soft-shelled crabs. Remove the gills and abdominal flaps. Wipe dry with paper towel. Cut each in half. Add 1/2 an egg white and some potato starch. Mix well.
2. Rinse the split mung beans. Soak in water till soft. Steam for 30 minutes until tender. Set aside.
3. Heat wok and add oil. Deep-fry the soft-shelled crabs until cooked through. Drain and save on a serving plate.
4. Heat wok and add oil. Stir-fry onion until fragrant. Add tomato and toss well. Put in finely diced pineapple and curry paste. Stir until fragrant. Add split mung beans, diced pineapple and peanut butter. Cook over low heat until thick. Pour in coconut milk and evaporated milk at last. Stir in the whisked egg and mix well. Pour the mixture over the soft-shelled crabs. Garnish with coriander. Serve.

必學不敗竅門

- 炸食物後的熱油，下一片薑再炸一會，可以去除油的肉腥味。
- After deep-frying any food in the oil, you may put in a slice of ginger and deep-fry it briefly. That would remove the gamey or fishy taste in the oil.

COOKING TIPS

海參
sea cucumber

釀脆皮海參
Deep-fried stuffed sea cucumber

材料

- 急凍日本刺參 8 條（已浸發）
- 熟鹹蛋黃 2 隻
- 蝦肉 8 隻
- 墨魚膠 200 克
- 椒鹽適量

Ingredients

- 8 frozen Japanese spiky sea cucumbers (rehydrated)
- 2 salted egg yolks (steamed)
- 8 shelled shrimps
- 200 g minced cuttlefish
- peppered salt

生粉脆漿料

- 生粉 5 湯匙
- 水 100 毫升
- 雞蛋 1 隻

Deep-frying batter

- 5 tbsp potato starch
- 100 ml water
- 1 egg

蘸醬

- 泰式雞醬 2 湯匙
- 梅子醬 2 湯匙

Dipping sauce

- 2 tbsp Thai sweet chilli sauce for fried chicken
- 2 tbsp plum sauce

做法

1. 刺參去腸，洗淨，瀝乾水分，撲上生粉，備用。
2. 熟鹹蛋黃切細條；蝦肉切長粒，備用。
3. 泰式雞醬與梅子醬拌勻，作蘸醬用。
4. 生粉加水拌勻，靜置一會，倒掉表層水分，留底層濕生粉，混入雞蛋拌勻成脆漿，備用。
5. 刺參內撲上生粉，依次釀入鹹蛋黃、蝦粒及墨魚膠（先撲生粉），刺參面再撲上生粉，沾上脆漿料，下油鑊炸至金黃色，最後灑上椒鹽，伴蘸醬同食。

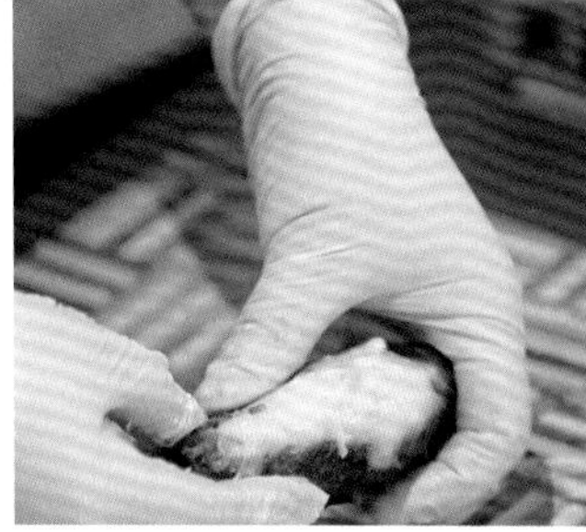

Method

1. Remove the guts of the sea cucumbers. Rinse and drain well. Coat them in potato starch.
2. Cut the salted egg yolks into small strips. Set aside. Cut the shrimps into elongated dices.
3. Mix the Thai sweet chilli sauce and plum sauce. Set aside.
4. To make the deep-frying batter, add water to potato starch. Let it rest for a while. Decant the water on top and use only the bottom layer of paste. Stir in an egg and mix well.
5. Coat the insides of the sea cucumbers with potato starch. Then stuff them in this particular order - salted egg yolk, diced shrimps and minced cuttlefish coated in potato starch. Coat the sea cucumber in potato starch again. Dip them into the deep-frying batter from step 4. Deep-fry in oil until golden. Sprinkle with peppered salt and serve with the dipping sauce on the side.

必學不敗竅門

- 用生粉、清水及雞蛋拌勻成脆漿料，令炸物有香脆的效果。
- The deep-frying batter made with potato starch, water and egg gives the food a crispy crust after deep-fried.

虎蝦
tiger prawn

沙律大蝦球
Fried tiger prawns in wasabi dressing

材料

- 虎蝦 400 克
- 蛋白 1/2 隻
- 生粉 2 茶匙

Ingredients

- 400 g tiger prawns
- 1/2 egg white
- 2 tsp potato starch

Wasabi 沙律醬

- 蟹子 1 湯匙
- 日式蛋黃醬 5 湯匙
- 煉奶 1 茶匙
- 泰式雞醬 1 湯匙
- 日式青芥末（wasabi）1 湯匙

Wasabi dressing

- 1 tbsp tobikko (flying fish roe)
- 5 tbsp Japanese style mayonnaise
- 1 tsp condensed milk
- 1 tbsp Thai sweet chilli sauce for chicken
- 1 tbsp wasabi (Japanese horseradish)

做法

示範短片

❶ 蝦解凍，洗淨後吸乾水分，在蝦背用刀劃一下及去腸，以少許鹽、胡椒粉略醃，加入蛋白及生粉。
❷ 將 Wasabi 沙律醬拌勻，備用。
❸ 燒熱油鑊，下蝦球炸至金黃，盛起。
❹ 燒熱另一鑊，下少許油，待油燒熱加入蝦球炒熱，再加入 Wasabi 沙律醬拌勻，上碟即成。

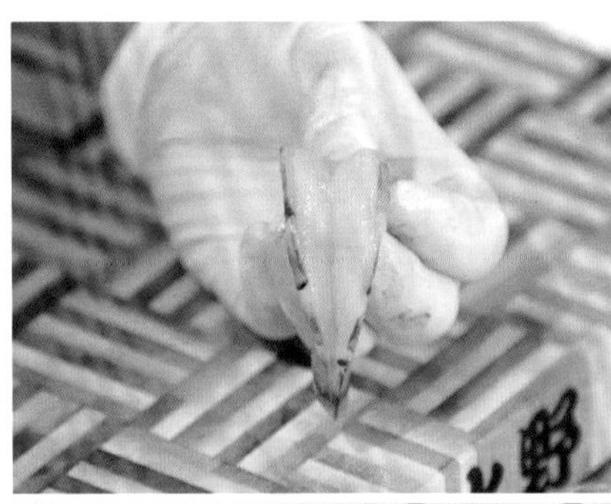

Method

❶ Thaw the tiger prawns. Rinse and wipe dry. Cut along the back and devein. Sprinkle with a pinch of salt and ground white pepper. Mix well. Stir in egg white and potato starch. Mix again.
❷ Mix all wasabi dressing ingredients together until well combined.
❸ Heat wok and add oil. Deep-fry the tiger prawns until golden. Drain.
❹ Heat another wok and add oil. Heat up the oil and put the prawns back in. Toss till heated through. Pour in wasabi dressing. Toss to coat evenly. Serve.

必學不敗竅門

- 虎蝦體型大，肉厚、少膏，適合炒、蒸。
- 想 Wasabi 沙律醬好味，秘訣是加入煉奶及泰式雞醬。
- Tiger prawns are large in sizes and meaty with little roe. They are great for stir-fried and steamed dishes.
- The trick to make the wasabi dressing taste divine - adding condensed milk and Thai sweet chilli sauce.

COOKING TIPS

生蠔
oyster

薑葱胡椒生蠔

Deep-fried oysters with ginger, spring onion and pepper

材料

美國桶蠔 1 桶
薑 80 克（切片）
葱 5-6 條（切段）
紅椒絲少許

Ingredients

- 1 tub U.S. oysters
- 80 g ginger (sliced)
- 5 to 6 spring onion (cut into short lengths)
- red chillies (shredded)

調味料

蠔油 2 茶匙
紹興酒 2 湯匙
鹽、糖各少許
胡椒粉 1 茶匙
黑椒碎少許

Seasoning

- 2 tsp oyster sauce
- 2 tbsp Shaoxing wine
- salt
- sugar
- 1 tsp ground white pepper
- ground black pepper

做法

1. 生蠔以生粉清洗乾淨，吸乾水分，放入沸水稍汆燙（先放入體型較大的），見表面收緊即可取出，吸乾水分後，再均勻撲上生粉。
2. 生蠔放入滾油，半煎炸至 8 成熟，盛起瀝油，備用。
3. 薑片拍鬆，放油鑊內爆香（勿不停翻動），待表面金黃，放入少許鹽炒勻，加入生蠔，灒入紹興酒，加入蠔油、鹽及糖調味，最後加入葱段、胡椒粉、黑椒碎炒勻，放入已燒熱砂鍋內，灒少許紹興酒，以紅椒絲裝飾，趁熱享用。

Method

1. Rub potato starch on oysters. Then rinse off the starch. Wipe dry. Blanch in boiling water briefly (the bigger ones go in first). When they start to tighten up on the outside, drain and wipe dry. Coat them in potato starch again.
2. Heat oil in a wok and put in the oysters. Cook in semi-deep frying manner until almost cooked through. Drain and set aside.
3. Crush the sliced ginger with the flat side of a knife. Heat wok and add oil. Stir-fry ginger until fragrant and golden (Do not keep stirring it). Add a pinch of salt and toss well. Put in the fried oysters. Drizzle with Shaoxing wine and oyster sauce. Season with salt and sugar. Lastly put in spring onion, ground white pepper and ground black pepper. Toss well. Transfer into a heated clay pot. Drizzle with a dash of Shaoxing wine. Garnish with red chillies. Serve hot.

必學不敗竅門

- 生蠔煮前以生粉輕捽可清除污物，用清水洗淨及印乾水分，以大沸水焯至蠔邊輕微收緊，有助定形。
- Rub the oysters in potato starch before rinsing. The starch clings to the dirt and sand so that they can be removed more easily. Then wipe them dry and blanch them in vigorously boiling water until they start to tighten up on the edges. This step would set their shape and plump them up.

雞翼
chicken wing

酥脆香蔥金沙雞翼

Fried chicken wings in salted egg yolk sauce with crispy shallot bits

材料

- 雞中翼 10 隻
- 鹹蛋黃 2 隻
- 牛油 10 克
- 油葱酥 2 湯匙
- 蛋白少許

Ingredients

- 10 chicken mid-joint wings
- 2 salted egg yolks
- 10 g butter
- 2 tbsp deep-fried shallot bits
- egg white

伴吃

- 酸子薑絲 40 克
- 青瓜絲 40 克

Garnish

- 40 g pickled young ginger (shredded)
- 40 g shredded cucumber

做法

1. 鹹蛋黃原個蒸熟，壓碎，備用。
2. 雞中翼洗淨，以鹽水浸泡片刻，於中間位置切一刀，加少許鹽、蛋白、胡椒粉拌勻，加入生粉約 1 湯匙拌勻，備用。
3. 大火燒熱油鑊，下雞翼，轉中小火煎香兩面至金黃。
4. 同步燒熱另一煎鑊，下牛油、熟鹹蛋黃碎及少許油，不停攪拌煮至起泡，下少許鹽及胡椒粉調味成金沙醬，放入雞翼大火炒勻，上碟，最後撒上油葱酥，伴以酸子薑絲、青瓜絲，趁熱享用。

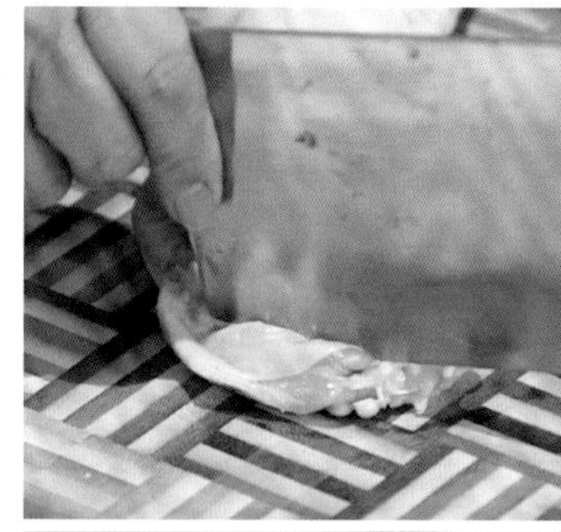

Method

1. Steam the salted egg yolks in whole until cooked. Mash them.
2. Rinse the chicken wings. Soak them in salted water briefly. Make a cut all the way through between the two bones. Add a pinch of salt, egg white and ground white pepper. Mix well. Stir in 1 tbsp of potato starch and mix well.
3. Heat wok over high heat. Add oil and heat it up. Shallow fry the chicken wings until both sides golden.
4. Meanwhile, heat another wok and add butter, salted egg yolks and some oil. Keep tossing until bubbly. Add a pinch of salt and ground white pepper. Put in the chicken wings and toss over high heat. Save on a serving plate. Sprinkle with deep-fried shallot bits. Garnish with pickled young ginger and shredded cucumber on the side of the plate. Serve hot.

必學不敗竅門

- 要煮出金沙的效果，必須使用牛油起鑊，下鹹蛋黃後可加油推至起泡。
- To make the salted egg yolk sauce creamy and aromatic, make sure you put in butter first after heating the wok. After adding the salted egg yolk, you may add some cooking oil and stir the mixture till bubbly.

海鮮 / 雞
seafood / chicken

帶子小炒皇

Hawker-style stir-fried chicken and assorted seafood

材料

Ingredients

- 急凍帶子 4 隻 — 4 frozen scallops
- 急凍虎蝦 10 隻 — 10 frozen tiger prawns
- 急凍雞扒 1 件 — 1 frozen boneless chicken thigh
- 韭菜花 40 克 — 40 g flowering chives
- 西芹 4 條 — 4 celery stems
- 鮮百合 1 個 — 1 fresh lily bulb
- 蝦米 10 隻 — 10 dried shrimps
- 蒜粒 1 湯匙 — 1 tbsp diced garlic
- 炸銀魚乾適量 — deep fried anchovies
- 炸腰果適量 — deep-fried cashew nuts
- 紅椒絲少許 — red chillies (shredded)

醃料

Marinade

- 糖 1/4 茶匙 — 1/4 tsp sugar
- 生粉 1/2 茶匙 — 1/2 tsp potato starch
- 生抽 1 湯匙 — 1 tbsp light soy sauce
- 麻油 1 茶匙 — 1 tsp sesame oil
- 胡椒粉少許 — ground white pepper

調味料

Seasoning

- 豆瓣醬 1-2 茶匙 — 1 to 2 tsp spicy bean sauce
- XO 醬 2 茶匙 — 2 tsp XO sauce
- 蝦米粉 1/2 茶匙 — 1/2 tsp ground dried shrimps
- 鹽 1/4 茶匙 — 1/4 tsp salt
- 紹興酒適量 — Shaoxing wine

必學不敗竅門

- 各材料煮熟的時間各有不同，所以分鑊炒熟後才一齊炒勻，效果最佳。
- As the cooking time is different for each ingredient, you must stir-fry them separately before assembling for the best result.

COOKING TIPS

做法

1. 帶子解凍，用沸水浸泡至半熟、定形，吸乾水分，撲上生粉，備用。
2. 虎蝦解凍、吸乾水分，加少許胡椒粉略醃，備用。
3. 西芹撕掉粗纖維部分，斜切小片；韭菜花切段；蝦米浸軟，瀝乾水分，備用。
4. 鮮百合洗淨，瀝乾，逐瓣剝開，去掉黑色部分，備用。
5. 雞扒解凍，洗淨、吸乾水分，切大粒，加入醃料拌勻略醃，備用。
6. 燒熱油鑊，下帶子及蝦，大火煎至兩面金黃色即盛起；帶子直切一開二。
7. 原鑊加油，放入雞肉粒爆炒，灒入紹興酒約 1 湯匙，大火快炒至 8 成熟，盛起。
8. 燒熱另一鑊，熱透後加入油，轉小火爆香蒜粒，轉大火下西芹片、雞肉粒、蝦米、鮮百合快炒，加鹽 1/4 茶匙、XO 醬炒勻。
9. 下韭菜花、蝦、帶子（去水），灒入紹興酒，加入豆瓣醬、蝦米粉調味，最後下紅椒絲炒勻，上碟，撒上炸銀魚乾及炸腰果即成。

Method

1. Thaw the scallops and soak them in hot water until half cooked and firmed up. Wipe dry and coat them in potato starch.
2. Thaw the tiger prawns. Wipe dry. Add a pinch of ground white pepper. Mix well.
3. Tear the tough veins off the celery stems. Slice diagonally. Set aside. Cut flowering chives into short lengths. Set aside. Soak dried shrimps in water until soft. Drain.
4. Rinse fresh lily bulb. Drain and break into scales. Cut off any dark bits.
5. Thaw the chicken thigh. Rinse and wipe dry. Dice coarsely. Add marinade and mix well.
6. Heat wok and add oil. Stir-fry scallops and prawns over high heat until both sides golden. Set aside. Cut each scallop in half lengthwise.
7. In the same wok, pour in some oil and stir-fry chicken. Drizzle with 1 tbsp of Shaoxing wine. Toss quickly until medium-well done. Set aside.
8. Heat another wok and add oil. Turn to low heat and stir-fry garlic until fragrant. Turn to high heat and put in celery, chicken, dried shrimps and lily bulb. Toss quickly. Add 1/4 tsp of salt and XO sauce. Stir to mix well.
9. Add flowering chives, prawns and scallops (drained). Drizzle with Shaoxing wine. Add spicy bean sauce and ground dried shrimps. Lastly, put in red chillies and toss well. Save on a serving plate. Sprinkled deep-fried anchovies and cashew nuts on top. Serve.

龍王翡翠燴魚肚

Braised fish tripe with pork, shrimps and cucumber

- 去除竹笙霉味，煮前可反覆浸水及擠乾，不斷換水至水變清澈透明為止。
- To remove the mouldy taste of bamboo fungus, soak it in water and squeeze dry repeatedly. After squeezing, soak it in fresh water. Repeat this step until the water runs clear.

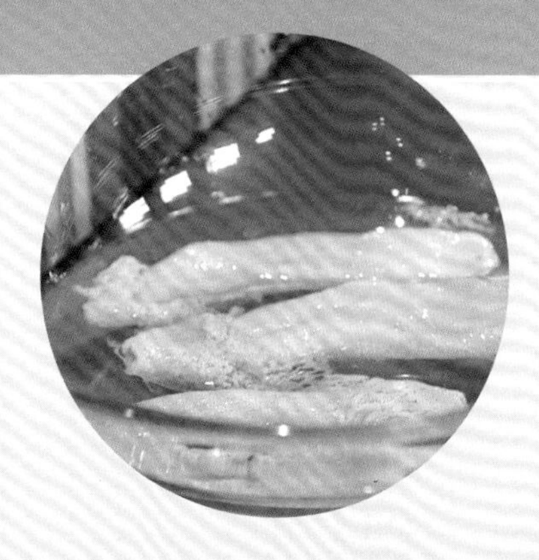

材料

鮮蝦 5 隻
大青瓜 1 條
砂爆魚肚 20 克
免治豬肉 200 克
鮮草菇 8 粒
竹笙 6 條
乾草菇 8 粒
大豆芽 80 克
上湯 200 毫升
蒜粒 1 湯匙
芹菜段少許

Ingredients

- 5 shrimps
- 1 large cucumber
- 20 g puffed fish tripe
- 200 g ground pork
- 8 fresh straw mushrooms
- 6 stems bamboo fungus
- 8 dried straw mushrooms
- 80 g soybean sprouts
- 200 ml stock
- 1 tbsp diced garlic
- Chinese celery (cut into short lengths)

做法

1. 免治豬肉加入少許鹽拌勻，備用。
2. 竹笙浸水後擠乾水分，重複 4-5 次至乾淨、無異味，剪掉底部備用。
3. 砂爆魚肚洗淨，泡水至軟身後擠乾，重複數次，以薑葱汆水，備用。
4. 青瓜相間地去皮、斜切厚片；大豆芽切半，備用
5. 蝦洗淨、吸乾水分，剪掉眼、觸鬚、足、頭尾尖角及胃（沙囊），備用。
6. 乾草菇洗淨、浸軟（水留用），備用。
7. 用白鑊大火烘乾大豆芽，盛起備用。
8. 燒熱油鑊，放入蝦煎香，灒入紹興酒，盛起備用。
9. 同一時間，燒熱另一油鑊，爆香蒜粒，放入免治豬肉炒香，下乾草菇連水快炒，加入大豆芽及青瓜炒勻，加入上湯、煎蝦、砂爆魚肚、竹笙及鮮草菇，以少許鹽、糖調味，加蓋再煮 2 分鐘，上碟，以芹菜段裝飾即成。

Method

1. Add a pinch of salt to the ground pork. Mix well.
2. Soak the bamboo fungus in water till soft. Drain and squeeze dry. Soak in fresh water again. Repeat this soaking and squeezing step 4 or 5 times until water runs clear and the bamboo fungus no longer smells stale. Cut off the root ends.
3. Rinse the fish tripe. Soak in water until soft. Squeeze dry. Repeat soaking and squeezing a few times. Blanch in boiling water with ginger and spring onion. Drain and set aside.
4. Peel the cucumber alternately to make stripes on the skin. Slice thickly at an angle. Set aside. Cut each soybean sprout in half. Set aside.
5. Rinse the shrimps and wipe dry. Cut off the eyes, antennae, legs, rostrums and the sac in the head with scissors.
6. Rinse the dried straw mushrooms. Soak in water till soft. Drain and set aside the soaking water for later use.
7. Fry the soybean sprouts in a dry wok. Toss until dry. Set aside.
8. Heat wok and add oil. Fry the shrimps until lightly browned. Drizzle with Shaoxing wine. Set aside.
9. Meanwhile, heat another wok and add oil. Stir-fry garlic until fragrant. Stir-fry ground pork until lightly browned. Add dried straw mushrooms and pour in the soaking water. Toss quickly. Add soybean sprouts and cucumber. Stir to mix well. Add stock, shrimps, fish tripe, bamboo fungus and fresh straw mushrooms. Toss and season with salt and sugar. Cover the lid and cook for 2 minutes. Save on a serving plate. Garnish with Chinese celery. Serve.

蟹粉
hairy crabmeat and roe

蟹粉河蝦薑葱煲仔飯

Clay pot rice with river shrimps, hairy crabmeat and roe

做法

1. 河蝦洗淨，吸乾水分，加入少許鹽、胡椒粉及生粉略醃，下蛋白拌勻，備用。
2. 大頭菜洗淨、浸泡，沖洗後瀝乾水分，切粒備用。
3. 米洗淨，放入小砂鍋內，加入凍水蓋至米面少許，開火煮沸。
4. 燒熱小油鑊，放入薑粒爆香，下少許鹽調味，加入大頭菜炒香，再放入蟹粉快炒，盛起備用。
5. 待飯煮滾（尚未乾水），放入部分蟹粉鋪面，加蓋，用較小火慢烘，期間不時轉動砂鍋令受熱均勻。
6. 同一時間，取小鍋燒油至暖，放入河蝦以小火炒熟，加入水 1 湯匙及少許鹽調味，下少許生粉芡炒勻，盛起。
7. 原鍋加入餘下蟹粉炒香，下少許水、鹽及胡椒粉調味，盛起放入河蝦面，與煲仔飯一同享用。
8. 處理飯面香葱料：燒熱小油鑊，放入紅葱粒爆香，下指天椒及少許鹽，關火後加入葱花炒勻，放進煲仔飯內，加入少許河蝦，趁熱享用時伴蟹醋份外滋味。

必學不敗竅門

- 炒蟹粉時，加入薑及大頭菜有提鮮之作用。
- When you stir-fry the hairy crabmeat and roe, adding ginger and salted kohlrabi helps accentuate their umami.

COOKING TIP

材料

Ingredients

- 蟹粉 200 克 — 200 g hairy crabmeat and roe
- 河蝦 160 克 — 160 g river shrimps
- 米 160 克 — 160 g rice
- 大頭菜 1 片 — 1 slice Cantonese salted kohlrabi
- 薑粒 3-4 湯匙 — 3 to 4 tbsp diced ginger
- 蛋白 1/2 隻 — 1/2 egg white

飯面香葱料

Spring onion topping

- 紅葱頭 2 湯匙（切粒） — 2 tbsp shallot (diced)
- 指天椒 1 隻（切圈） — 1 bird's eye chilli (cut into rings)
- 葱花 8-10 湯匙 — 8 to 10 tbsp finely chopped spring onion

Method

1. Rinse the shrimps and wipe dry. Add a pinch of salt, ground white pepper and potato starch. Mix well and leave them briefly. Add egg white and stir well.
2. Rinse the salted kohlrabi. Soak it in water. Rinse again and drain. Dice and set aside.
3. Rinse the rice and drain. Put into a clay pot. Add cold water to cover. Put it over the stove. Turn on medium-high heat. Bring to the boil.
4. Heat a small wok or pan. Add oil and stir-fry ginger until fragrant. Put in kohlrabi and toss until fragrant. Put in the hairy crabmeat and roe. Toss quickly and set aside.
5. When the rice mixture is boiling and you can still see some water over the rice, arrange some of the hairy crabmeat mixture from step 4 over the rice. Cover the lid and turn to low heat. Tilt and turn the pot around to heat the vertical side of the pot evenly.
6. Meanwhile, heat some oil in a small pan until warm. Stir-fry the shrimps over low heat until cooked through. Add 1 tbsp of water and a pinch of salt. Stir in a little potato starch thickening glaze. Toss and set aside.
7. In the same pan, stir-fry the remaining hairy crabmeat mixture until fragrant. Add some water, salt and ground white pepper. Transfer over the river shrimps. Set aside.
8. To make the spring onion topping, heat a small pan. Stir-fry shallot until fragrant. Add bird's eye chilli and a pinch of salt. Turn off the heat and stir in the spring onion. Put the mixture on top of the cooked rice. Then arrange the river shrimp and hairy crabmeat mixture on top. Serve hot with black vinegar on the side.

古法鹽焗雞

Old-fashioned salt-baked chicken

材料

- 雞 1 隻
- 粗鹽 4800 克（8 斤）
- 薑汁 4 湯匙
- 八角 6 茶匙（拍碎）
- 花椒 3 粒
- 紗紙 2 張（包雞用）

Ingredients

- 1 chicken
- 4.8 kg coarse salt
- 4 tbsp ginger juice
- 6 tsp star-anise (crushed)
- 3 Sichuan peppercorns
- 2 sheets mulberry paper (for wrapping the chicken)

做法

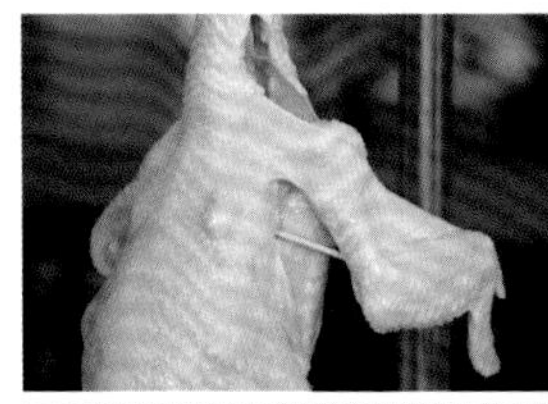

1. 雞洗淨、抹乾，取鹽 1 茶匙與薑汁混和，塗勻雞內外，放雪櫃醃一晚，備用。
2. 雞取出，充分抹乾，吊起風乾 3-4 小時（用牙籤撐起翼位），完成後用紗紙包封好（先掃油防黏），備用。
3. 將粗鹽分成兩份，分別放入兩個炒鑊，用大火不停炒香，待鹽熱透、冒煙，放入八角及花椒同炒，待鹽炒至微黃，放入雞（背朝天），再放另一半鹽蓋過雞，加蓋煮 10 分鐘，轉小火再焗 10 分鐘，最後關火焗 20 分鐘即成。

Method

1. Rinse the chicken and wipe dry. Mix 1 tsp of salt and ginger juice. Pour the mixture on the chicken and rub it over both the inside and the outside. Refrigerate overnight.
2. Take chicken out of the fridge. Wipe dry. Then hang the chicken in an airy spot for 3 to 4 hours. Prop the wings up with toothpicks so that the fold of skin is also exposed. Brush oil all over so that the skin won't stick to the mulberry paper. Then wrap it in mulberry paper.
3. Divide the salt into 2 equal parts. Put them in two dry woks. Heat over high heat and keep tossing continuously until fragrant. When the salt is giving off steam and heated through, put in the star-anise and Sichuan peppercorns. Toss until the salt turns lightly yellow. Put the paper-wrapped chicken in with the breast side down. Pour the salt in the other wok over the chicken. Cover the lid and cook for 10 minutes. Turn to low heat and cook for 10 minutes. Turn off the heat at last and leave the chicken in the covered wok for 20 more minutes. Serve.

必學不敗竅門

- 粗鹽必須炒至熱透才放入雞，利用熱力讓雞熟透。
- 紗紙可於紙紮店購買，或用煮食紙取代。
- The salt must be stir-fried till hot before putting the chicken in. That ensures the heat penetrates the chicken quickly and cooks it through.
- You can buy mulberry paper from traditional paper-ware stores. But you may also use parchment or baking paper instead.

牛柳邊
beef striploin

啫啫紫蘿牛肉
Sizzling beef and pineapple in clay pot

材料

- 新鮮牛柳邊 400 克
- 鮮菠蘿 1/4 個
- 鮮菠蘿粒 3 湯匙
- 新鮮子薑 80 克（切薄片）
- 醃酸薑片 40 克
- 青、紅燈籠椒各 1/4 個（切片）
- 薑 4 片
- 糖醋汁 2 湯匙
- 鮮醬油 1 茶匙

醃料

- 糖 1/4 茶匙
- 油 1/2 湯匙
- 生粉 1 茶匙
- 生抽 1/2 湯匙
- 胡椒粉少許

Ingredients

- 400 g fresh beef striploin
- 1/4 fresh pineapple
- 3 tbsp diced fresh pineapple
- 80 g fresh young ginger (sliced thinly)
- 40 g pickled ginger (sliced)
- 1/4 green bell pepper (sliced)
- 1/4 red bell pepper (sliced)
- 4 slices ginger
- 2 tbsp sweet and sour sauce
- 1 tsp Maggi seasoning

Marinade

- 1/4 tsp sugar
- 1/2 tbsp oil
- 1 tsp potato starch
- 1/2 tbsp light soy sauce
- ground white pepper

做法

1. 鮮菠蘿去皮，切件，取 2-3 件切成碎粒，備用。
2. 新鮮牛柳邊切片，加入醃料拌勻，備用。
3. 砂鍋燒熱，放入菠蘿片，加蓋，用小火烘香兩面，備用。
4. 燒熱油鑊，爆香青、紅椒，盛起備用。
5. 原鑊加入薑片爆香，放入牛柳片煎至半熟，盛起備用。
6. 原鑊加入子薑片爆香，加入鮮菠蘿碎粒、青紅椒、醃酸薑片及牛柳片快炒，最後加入糖醋汁及鮮醬油炒勻，倒入盛有菠蘿片的砂鍋拌勻，趁熱享用。

Method

1. Peel the fresh pineapple and slice it. Then finely dice 2 to 3 slices. Set aside the rest.
2. Slice the beef. Add marinade and mix well.
3. Heat a clay pot. Put in the sliced pineapple. Cover the lid. Cook over low heat until both sides browned.
4. Heat wok and add oil. Stir-fry bell peppers until fragrant. Set aside.
5. In the same wok, stir-fry ginger until fragrant. Add beef and fry until half-cooked. Set aside.
6. In the same wok, stir-fry young ginger until fragrant. Put in the diced pineapple, bell peppers, pickled ginger and beef. Season with Maggi, sweet and sour sauce. Toss quickly. Transfer the mixture into the clay pot over the sliced pineapple. Toss well and serve the whole pot.

牛柳邊比牛柳有較多脂肪，牛味更濃。
It's advisable to ask the butcher for the fatty rim of beef striploin, which has stronger meaty flavours and more marbling.

必學不敗竅門

- 菠蘿煮前先用白鑊烘乾，味道更甜、更多汁。
- Before cooking pineapple, fry it in a dry wok first. That would make the pineapple sweeter and juicier.

COOKING TIPS

牛仔骨
beef short ribs

洋葱紅酒牛仔骨

Braised beef short ribs in onion red wine

材料

- 牛仔骨（英式切法）3 件（約 1 公斤）
- 紫洋葱 1 個
- 甘筍 1/2 條
- 薑 3 片
- 蒜肉 2 瓣（連皮）
- 紅葱頭 4 粒（原粒）
- 芹菜 1 棵
- 芫茜 2 棵
- 小茴香籽 1 茶匙
- 上湯 400 毫升
- 麵粉 5 湯匙
- 黑椒碎少許
- 洋葱紅酒 2-3 湯匙
- 鮮醬油 1 茶匙

Ingredients

- 3 pieces beef short ribs (English-cut, about 1 kg)
- 1 red onion
- 1/2 carrot
- 3 slices ginger
- 2 cloves garlic (skin on)
- 4 shallots
- 1 sprig Chinese celery
- 2 sprigs coriander
- 1 tsp cumin seeds
- 400 ml stock
- 5 tbsp flour
- ground black pepper
- 2 to 3 tbsp onion red wine
- 1 tsp Maggi seasoning

浸牛仔骨料

- 洋葱紅酒 700 毫升
- 香葉 2 片
- 八角 1 粒
- 小茴香籽 1/4 茶匙

Marinade

- 700 ml onion red wine
- 2 bay leaves
- 1 whole star-anise
- 1/4 tsp cumin seeds

做法

1. 牛仔骨放大盤內，倒入洋葱紅酒浸過牛仔骨面，放入香葉、八角及小茴香籽封好，放雪櫃浸泡一晚，備用。
2. 甘筍去皮，滾刀切角；芹菜梗切粒、葉切碎；芫茜切根留用，其餘切碎；紫洋葱切片，備用。
3. 燒熱油鑊，放入甘筍、薑片、蒜肉、紫洋葱、紅葱頭、芹菜、芫茜根、小茴香籽爆香，倒入半份上湯，加蓋煮滾。
4. 牛仔骨均勻地撲上麵粉，放入油鑊煎香，煎至每面微焦，取出放入③內，加入餘下上湯、少許小茴香籽及黑椒碎，蓋好小火炆 45 分鐘。
5. 完成後將牛仔骨取出，放另一鑊內，隔渣取汁，再加入洋葱紅酒及鮮醬油同煮滾，加入生粉芡煮至稠，上碟後以芫茜裝飾即成。

Method

1. Put the beef into a big bowl or tray. Pour in onion red wine to cover. Add bay leaves, star-anise and cumin. Cover in cling film. Refrigerate overnight.
2. Peel carrot and cut into random wedges while rolling it on the chopping board. Set aside. Dice the Chinese celery stems. Finely chop the leaves. Set aside. Cut off the coriander roots and set aside. Finely chop the stems and leaves. Set aside. Slice the red onion.
3. Heat wok and add oil. Stir-fry carrot, ginger, garlic, red onion, shallots, Chinese celery, coriander roots and cumin seeds until fragrant. Pour in half of the stock. Cover the lid and bring to the boil.
4. Coat the beef evenly in flour. Fry in oil until lightly browned on all sides. Transfer the beef into the vegetable mixture from step 3. Pour in the remaining stock. Add cumin seeds and ground black pepper. Cover the lid and simmer over low heat for 45 minutes.
5. Transfer the beef into another wok. Strain the cooking stock into this wok. Add onion red wine and Maggi seasoning. Bring to the boil. Stir in potato starch thickening glaze. Cook until it thickens. Save on a serving dish. Garnish with coriander. Serve.

必學不敗竅門

- 用洋葱紅酒先將牛仔骨醃過夜，味道提升，更加入味。
- Marinating the beef in onion red wine overnight helps elevate the flavours and ensures infusion of the seasoning.

大尾魷
oval squid

叉燒汁煮荷包魷

Braised oval squid in barbecue sauce

材料

- 荷包魷（大尾魷）1 隻
- 甘草欖角 2 湯匙（做法參考 p.15）
- 叉燒醬 2 湯匙
- 蜂蜜 1 湯匙
- 紅麴米 1 湯匙
- 玫瑰露 1 茶匙
- 香茅 1 條
- 紅葱頭 2 粒（切片）
- 炒香芝麻適量

Ingredients

- 1 oval squid
- 2 tbsp preserved black olives (method refer to p.15)
- 2 tbsp Chinese barbecue sauce
- 1 tbsp honey
- 1 tbsp red yeast rice
- 1 tsp Chinese rose wine
- 1 stem lemongrass
- 2 shallots (sliced)
- toasted sesames

做法

1. 將荷包魷身及觸鬚分開，去皮及內臟，保留魷魚骨，洗淨及吸乾水分，均勻地撲上生粉，備用。
2. 香茅切段、拍扁，備用。
3. 紅麴米蒸熟後浸水，隔米，水留用。
4. 熱鑊下油，加入香茅段爆香後，夾起香茅段，下荷包魷稍煎香，再下紅葱頭、甘草欖角、叉燒醬、適量紅麴米水，煮至收汁（期間翻轉另一面煮至兩面均勻），最後加入蜂蜜，灒入玫瑰露，取走魷魚骨後切件上碟，淋上汁，撒上芝麻即成。

示範短片

Method

1. Separate the mantle and the tentacles of the squid. Peel off the purple skin and remove all innards. Keep the bone inside. Rinse and wipe dry. Coat it in potato starch evenly.
2. Cut lemongrass into short lengths. Bruise them with the back of a knife.
3. Steam the red yeast rice. Soak it in water. Drain and set aside the soaking water.
4. Heat wok and add oil. Stir-fry lemongrass until fragrant. Remove the lemongrass. Put in the squid to fry until lightly browned. Then add shallot, preserved black olive, barbecue sauce, and some of the water from soaking red yeast rice. Cook until the liquid reduced and flip the squid once or twice throughout the process. Add honey and Chinese rose wine at last. Transfer the squid onto a serving plate. Remove the bone inside. Slice it. Dribble the sauce over. Sprinkle with sesames. Serve.

必學不敗竅門

- 煮荷包魷時保留魷魚骨在內，煮熟後才取走，能保持魷魚完整。
- When you cook the oval squid, it's advisable to keep the bone inside as it keep the squid in whole throughout the cooking process. Remove the bone after the squid is cooked through.

COOKING TIPS

冬瓜

winter melon

迷你海鮮冬瓜盅

Seafood soup in whole mini winter melon

必學不敗竅門

- 因鹽有提鮮作用，冬瓜於蒸前抹上鹽，令冬瓜味道更香、更突出。
- Rub salt on the inside of the winter melon before steaming it. The salt would accentuate the flavours and aromas of the winter melon.

COOKING TIP

材料

- 迷你冬瓜 1 個
- 勝瓜 40 克（去皮、切粒）
- 燒鴨腿 1 隻（去骨、切絲）
- 脢頭肉 80 克（切粒）
- 海蝦 4-5 隻
- 蟹柳 2 條（拆絲）
- 瑤柱 6 粒
- 金華火腿 40 克（切片）
- 乾草菇 8 粒
- 夜香花 40 克
- 新鮮蓮子 10 粒（去芯）
- 鮮草菇 8 粒
- 上湯 500 毫升
- 大地魚粉 1 茶匙

Ingredients

- 1 mini winter melon
- 40 g angled loofah (peeled, diced)
- 1 store-bought roast duck leg (de-boned, shredded)
- 80 g pork shoulder butt (diced)
- 4 to 5 marine shrimps
- 2 imitation crabsticks (torn into shreds)
- 6 dried scallops
- 40 g Jinhua ham (sliced)
- 8 dried straw mushrooms
- 40 g floral buds of night willow herb
- 10 fresh lotus seeds (cored)
- 8 fresh straw mushrooms
- 500 ml stock
- 1 tsp ground dried plaice

做法

1. 迷你冬瓜洗淨，從瓜頂切開 1/5 左右（留用），瓜邊用小刀劃成鋸齒花（可省略），以刀尖在瓜肉中央劃「十」字，去瓤及刮成中空，用鹽塗抹冬瓜內部，隔水蒸 10 分鐘。
2. 瑤柱及乾草菇洗淨、浸軟，瑤柱水留用。
3. 海蝦洗淨，去頭、去殼、去腸，留尾。
4. 金華火腿、脢頭肉粒、海蝦、乾草菇、蟹柳汆水備用。
5. 上湯放鍋內加熱，加入金華火腿、瑤柱、瑤柱水、肉粒、乾草菇、鮮草菇、鮮蓮子煮 30 分鐘，再加進部分火鴨絲。
6. 將上湯料放入冬瓜內，繼續蒸約 30 分鐘。
7. 最後放入勝瓜粒，海蝦、蟹柳及少量夜香花放冬瓜盅邊作點綴，並把部分夜香花、火鴨絲放進冬瓜盅，加入大地魚粉調味，以切出瓜頂作底座，放上冬瓜盅即成。

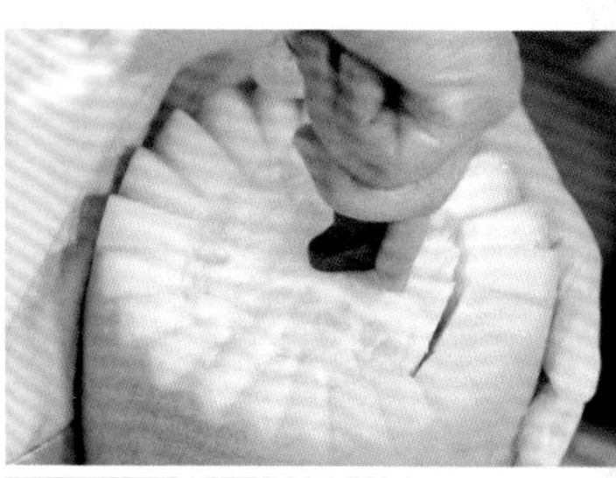

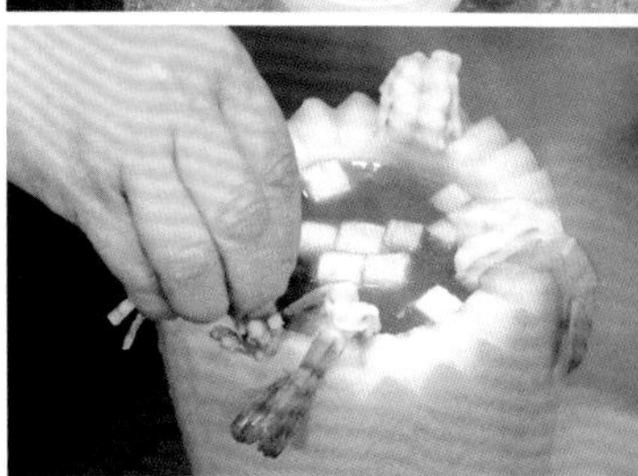

Method

1. Rinse the mini winter melon. Cut off 1/5 from the top and set aside. Then use a paring knife to cut out little triangles from the rim of the bottom 4/5 so that it looks zigzag. (This step is purely cosmetic, and therefore optional.) Cut a "X" at the centre of the flesh. Scoop out the seeds and part of the flesh. Rub salt on the inside of the winter melon. Steam for 10 minutes.
2. Rinse dried scallops and dried straw mushrooms. Soak them in water till soft. Save the water from soaking dried scallops for later use.
3. Rinse the shrimps. Remove the head and shell them. Devein, but keep the tails intact.
4. Blanch Jinhua ham, pork shoulder butt, marine shrimps, dried straw mushrooms and imitation crabsticks in boiling water. Drain.
5. Boil the stock in a pot. Put in Jinhua ham, dried scallops and the soaking water, pork, dried and fresh straw mushrooms, and lotus seeds. Boil for 30 minutes. Put in the shredded roast duck.
6. Transfer the stock and the solid ingredients into the mini winter melon. Keep on steaming for 30 minutes.
7. Lastly arrange diced angled loofah, shrimps, imitation crabsticks and some night willow herb along the rim of the winter melon as garnish. Put some of the night willow herb and roast duck into the stock. Season with ground dried plaice. Cut the cut-off top of the winter melon into a base for the steamed winter melon to sit on. Serve.

← 迷你冬瓜 5-8 月是當造期，挑選體形平均及有小絨毛的，代表最新鮮。

Mini winter melon is in season from May to August each year. Pick those with an even girth from top to bottom, with fine hair on the skin, which means the winter melon is fresh.

豬肉 / 雞蛋
pork / egg

海龍皇蒸肉餅伴水蛋

Steamed pork patty and egg custard with dried scallops

必學不敗竅門

- 蒸水蛋時間控制最重要，按示範的份量蒸約 4 分鐘、再焗 3 分鐘，效果最理想。
- 蛋液必須用隔篩過濾，去掉蛋筋，蒸出來的蛋更滑溜。
- 肉餅鋪入碟後，輕輕在肉面按出空隙，蒸出來的肉餅更滑、更有口感。
- The key to velvety and jiggly egg custard is the steaming time. For the amounts listed in this recipe, steam it for 4 minutes and leave it in the steamer or wok for 3 minutes after turning off the heat, to achieve the best results.
- Make sure you pass the whisked eggs through a wire mesh to remove any lumps or the stringy egg white. The custard would be silkier that way.
- After you put the chopped pork mixture on a steaming plate, gently press the surface to create small indentations. The pork patty would turn out more velvety, with slightly more complex texture.

COOKING TIPS

肉餅材料

- 五花腩 250 克
- 豬板筋 50 克
- 冬菇 3 朵（浸水）
- 瑤柱 4 粒
- 葱花 3-5 湯匙
- 紅尖椒粒少許（裝飾）
- 熟蝦 2 隻（裝飾）

Ingredients for pork patty

- 250 g pork belly
- 50 g pork silver skin
- 3 dried shiitake mushrooms (soaked in water till soft)
- 4 dried scallops
- 3 to 5 tbsp finely chopped spring onion
- diced red chilli (as garnish)
- 2 cooked shrimps (as garnish)

調味料

- 鹽 1/2 茶匙
- 糖少許
- 生粉 1.5 茶匙
- 上湯 1-2 湯匙

Seasoning

- 1/2 tsp salt
- sugar
- 1.5 tsp potato starch
- 1 to 2 tbsp stock

水蛋材料

- 雞蛋 3 隻
- 上湯 250 毫升

Egg custard

- 3 eggs
- 250 ml stock

做法

示範短片

1. 冬菇洗淨，浸軟、切粒；瑤柱洗淨，浸軟，撕成絲，瑤柱水留用。
2. 豬板筋洗淨、切粒備用（切前可略汆燙以防「滑刀」）。
3. 五花腩洗淨、切粒，剁成肉碎，加入豬板筋粒搓勻。
4. 肉碎加入冬菇粒、瑤柱絲及瑤柱水，加入鹽、糖及生粉拌勻調味，再加入上湯搓勻及撻至起膠，放大蒸碟隔水大火蒸 15 分鐘。
5. 雞蛋拂勻，加少許鹽及上湯（1 份雞蛋：1.5 份上湯）拌勻，用篩隔走泡沫及雜質，盛起蛋液備用。
6. 把蒸好的肉汁加入蛋液內，輕力拌勻，將蛋液倒入肉餅邊，用保鮮紙封好，中小火隔水蒸 4 分鐘，熄火後焗 3 分鐘至蛋剛熟。
7. 燒熱鑊，熄火後下油，爆香紅椒粒，盛起備用。再下葱花爆香，加少許鹽盛起，放進肉餅邊，最後放上熟蝦及紅椒粒裝飾即成。

Method

1. Rinse the shiitake mushrooms. Soak them in water till soft. Cut off the stems and dice them. Set aside. Soak dried scallops in water till soft. Drain and set aside the soaking water. Tear into shreds.
2. Rinse the pork silver skin. Dice and set aside. (Optionally, blanch it in boiling water first so that it is easier to dice.)
3. Rinse the pork belly. Dice it and then finely chop it. Add the pork silver skin from step 2. Mix in a mixing bowl.
4. Add shiitake mushrooms, dried scallops and the soaking water to the pork mixture. Season with salt and sugar. Add potato starch. Mix well. Stir in the stock and mix well. Lift the mixture off the bowl and slap it forcefully back in a few times until sticky. Transfer the mixture onto a large steaming plate. Shape it into a round patty. Steam over high heat for 15 minutes. Drain any liquid on the plate and save for later use.
5. Whisk the eggs and add a pinch of salt and stock (1 part eggs to 1.5 parts stock). Mix well. Pass the mixture through a wire mesh to remove any lumps or foam.
6. Add the cooled pork juices from step 4 to the whisked eggs. Stir gently. Then pour the whisked eggs into the steaming plate around the pork patty. Cover in cling film and steam over medium-low heat for 4 minutes. Turn off the heat and leave it in the steamer or wok for 3 more minutes until the egg just sets.
7. Heat a wok. Turn off the heat and add oil. Stir-fry the red chilli until fragrant. Set aside. Stir-fry the spring onion until fragrant. Add a pinch of salt and set aside. Arrange the spring onion around the pork patty. Put cooked shrimps and red chilli over the pork. Serve.

紅花蟹

red swimmer crab

半煎煮冬瓜蟹

Braised crab with winter melon

材料

紅花蟹 1 隻（約 600 克）
冬瓜 200 克
鮑貝 10 粒
上海白年糕 8 條（切幼條）
大頭菜 2 片
薑 8 片（拍扁）
葱 1 棵（切段）
芹菜 1 棵（切段）
紹興酒 2 湯匙
生抽 2 茶匙
蝦米粉 1 茶匙
鹽 1/4 茶匙
糖 1/4 茶匙

Ingredients

- 1 red swimmer crab (about 600 g)
- 200 g winter melon
- 10 Pacific clams
- 8 strips Shanghainese rice cakes (cut into fine strips)
- 2 slices Cantonese salted kohlrabi
- 8 slices ginger (crushed)
- 1 spring onion (cut into short lengths)
- 1 sprig Chinese celery (cut into short lengths)
- 2 tbsp Shaoxing wine
- 2 tsp light soy sauce
- 1 tsp ground dried shrimps
- 1/4 tsp salt
- 1/4 tsp sugar

做法

❶ 大頭菜重複浸洗數次去掉多餘鹹味，搾乾水分後切絲。
❷ 年糕切成約 4 厘米條，用熱水浸軟，備用。
❸ 冬瓜洗淨去皮，切成約 4 厘米粗條，放熱水加蓋略煮，熄火後加入年糕同焗。
❹ 紅花蟹洗淨，蟹身斬成 4 件，蟹鉗先用刀背拍碎，撲上適量生粉，備用。
❺ 熱鑊下油，先爆香薑片，下蟹件煎香，加蓋稍焗；反轉蟹件再煎香另一面，加蓋稍焗。
❻ 焗至蟹半熟，灒入紹興酒、生抽，加入大頭菜、冬瓜條及年糕，下水適量及少許鹽、糖調味，再加入鮑貝，加蓋炆至水分略收乾。
❼ 最後加入少許蝦米粉，下葱段、芹菜兜勻，上碟即成。

Method

❶ Soak and drain the salted kohlrabi in fresh water repeatedly to make it less salty. Squeeze dry and finely shred it.
❷ Cut the Shanghainese rice cakes into strips about 4 cm long. Soak in hot water briefly. Drain.
❸ Rinse and peel the winter melon. Cut into thick strips about 4 cm long. Boil a pot of water and put in the winter melon. Cover the lid and cook briefly. Turn off the heat and add the Shanghainese rice cakes. Cover the lid again. Drain right before using.
❹ Rinse and dress the crab. Cut the body into 4 pieces. Crack the pincers with the back of a knife. Coat the crab pieces in potato starch.
❺ Heat wok and add oil. Stir-fry ginger until fragrant. Put in the crab pieces and fry until lightly browned. Cover the lid and cook briefly. Flip the crab pieces to fry the other side. Cover the lid and cook further.
❻ When the crab is half-cooked, drizzle with Shaoxing wine and light soy sauce. Put in the salted kohlrabi, winter melon and rice cakes. Add some water and a pinch of salt and sugar. Put in the Pacific clams. Cover the lid and cook until the liquid reduces.
❼ Add ground dried shrimps at last. Sprinkle with spring onion and Chinese celery. Toss and serve.

必學不敗竅門

➔ 紅花蟹
Red swimmer crab

- 用刀拍蟹鉗前宜用毛巾蓋着，否則蟹殼會四濺。
- 冬瓜放入沸水後蓋上鑊蓋焗片刻，冬瓜才夠腍身；冬瓜索滿蟹的鮮香味，是最搶手的配菜。
- Before you crack the pincers, wrap them in a towel first. Otherwise, the shell may crack into tiny bits and splatter everywhere.
- After putting the winter melon strips into boiling water, make sure you cover the lid and cook briefly. Otherwise, the winter melon may not be cooked through and soft enough. In this recipe, the winter melon is the hero, as it sucks up the juices from the crab.

COOKING TIPS

海鮮

seafood

大芥菜海鮮煲

Seafood and mustard greens in clay pot

材料

- 花蟹 1 隻
- 花蛤 300 克
- 芥菜（包心芥菜）1 個
- 鹹酸菜 1/2 棵
- 黃豆 100 克
- 薑 3 片
- 上湯 1 公升
- 豬骨 500 克
- 金華火腿 80 克

Ingredients

- 1 swimmer crab
- 300 g Venus clams
- 1 head mustard greens
- 1/2 head pickled mustard greens
- 100 g soybeans
- 3 slices ginger
- 1 litre stock
- 500 g pork bones
- 80 g Jinhua ham

做法

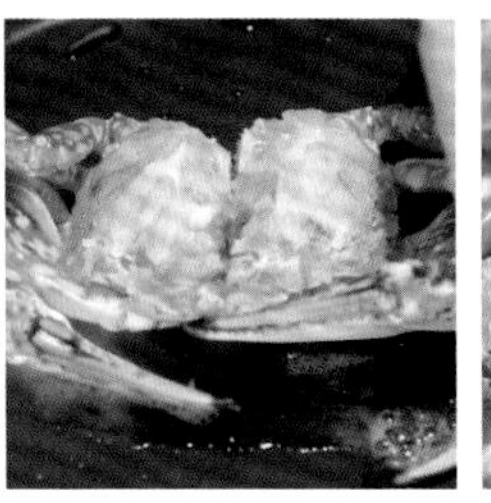

1. 黃豆洗淨先浸泡過夜（至少 2 小時），備用。
2. 芥菜洗淨、切件；鹹酸菜洗淨後浸泡，切件。
3. 豬骨以粗鹽 1/2 湯匙先醃過夜，用前先汆水。
4. 金華火腿先汆水後切粒，備用。
5. 花蛤連海水放雪櫃冷藏 4 小時，取出後宜棄去沒開殼的花蛤，沖洗乾淨。
6. 花蟹清洗好，去掉鰓、胃、腸，蟹身斬開 2 件，吸乾水分，備用。
7. 上湯倒入鍋內，加入芥菜、鹹酸菜、黃豆、薑片、豬骨及半份金華火腿，煮 15-20 分鐘。
8. 熱鑊下油，下蟹煎香，蟹蓋、蟹身有膏一面朝上，蓋好煎煮至半熟，取出放入鍋內，加入花蛤及餘下的金華火腿，蓋好煮至花蛤全開口即成。

Method

1. Soak soybeans in water for at least 2 hours or overnight. Drain.
2. Rinse fresh mustard greens and cut into pieces. Set aside. Rinse and soak the pickled mustard greens in water. Cut into pieces.
3. Rub 1/2 tbsp of coarse salt over the pork bones. Leave them overnight. Blanch in boiling water before using. Drain.
4. Blanch Jinhua ham in boiling water. Drain and dice it.
5. Put the clams in salted water. Refrigerate for 4 hours. Discard any clams that don't open their shells. Rinse well.
6. Rinse and dress the crab. Remove the gills, the sandy sac and the digestive tract. Chop the body in half. Wipe dry.
7. Boil stock in a pot. Put in the fresh and pickled mustard greens, soybeans, ginger, pork bones and half of the Jinhua ham. Cook for 15 to 20 minutes.
8. Heat wok and add oil. Fry the crab until fragrant. Then flip it so that the side with roe faces upward. Cover the lid and cook until half-done. Transfer into a clay pot. Add the clams and the rest of the Jinhua ham. Cook until all clams open. Serve the whole pot.

必學不敗竅門

- 芥菜不宜切得太細塊，否則煮的時候很容易爛。
- 買花蛤時可請檔主給些海水，回家後將花蛤浸在海水內，放入雪櫃 4 小時後，花蛤自動會將沙吐出。
- Cut the mustard greens into chunky pieces. Otherwise, they tend to break down and turn mushy after cooked.
- When you shop for live clams, you may ask the fishmonger to give you some sea water. Then put them in the sea water and refrigerate for 4 hours for them to spit out the sand.

COOKING TIPS

龍躉
giant grouper

椰香豆漿龍躉球

Giant grouper fillet in coconut soymilk

必學不敗竅門

- 煮豆漿要用文火，以避免豆漿滾瀉及豆漿面凝成豆皮。
- 蛋白倒入豆腐時不要攪勻，讓它慢慢凝結就可以，否則蛋白會散掉。
- Always cook soymilk over low heat, as it tends to boil over easily and a skin on form on the surface if the heat is too strong.
- When you pour the egg white into the soymilk, do not stir it. Just let it coagulate slowly. Otherwise, the egg white will be broken down into fine strands.

COOKING TIP

材料

- 龍躉腩 400 克
- 蜆肉 200 克
- 椰青水 200 毫升
- 無糖濃豆漿 500 毫升
- 白靈菇 1/2 個
- 鮮腐皮 80 克
- 蛋白 1 隻
- 杞子 1 茶匙
- 芹菜粒 1 茶匙

Ingredients

- 400 g giant grouper belly
- 200 g shelled clams
- 200 ml young coconut water
- 500 ml unsweetened concentrated soymilk
- 1/2 pleurotus nebrodensis mushroom
- 80 g fresh beancurd skin
- 1 egg white
- 1 tsp dried goji berries
- 1 tsp diced Chinese celery

做法

1. 白靈菇抹乾淨，切片。杞子洗淨浸軟，汆水備用。
2. 蜆肉洗淨先汆水，備用。鮮腐皮洗淨、切件。
3. 龍躉洗淨切片，以少許胡椒粉、生粉拌勻，再下蛋白少許及鹽適量拌勻略醃，備用。
4. 將白靈菇放油鑊煎至金黃，盛起備用。
5. 將龍躉片放熱油鑊煎香，盛起備用（亦可省卻此步驟，直接放豆漿內煮熟）。
6. 豆漿倒入鍋內以小火邊攪拌邊煮熱，依次下白靈菇、鮮腐皮、椰青水、龍躉片、蜆肉煮滾（注意勿大火，免滾瀉），最後倒入餘下蛋白，即可上碟，放上杞子、芹菜粒伴碟即成。

Method

1. Wipe down the pleurotus nebrodensis mushroom and slice it. Set aside. Rinse and soak goji berries in water until soft. Blanch in boiling water. Drain.
2. Rinse and blanch shelled clams in boiling water. Drain and set aside. Rinse and cut fresh beancurd skin into pieces.
3. Rinse and slice the fish. Sprinkle with a pinch of ground white pepper and potato starch. Mix well. Add some whisked egg white and salt. Mix and leave it briefly.
4. Heat wok and add oil. Fry the pleurotus nebrodensis mushroom in some oil until golden on both sides. Set aside.
5. In the same wok, fry the fish until both sides golden. Set aside. (This step is optional. You may put the raw fish directly into the soymilk if you prefer.)
6. Pour soymilk into a pot. Bring to the boil over low heat while stirring continuously. Then put in pleurotus nebrodensis mushroom, fresh beancurd skin, young coconut water, sliced grouper belly and shelled clams. Bring to the boil. (Watch the pot always as soymilk tends to boil over easily.) Pour in some egg white at last. Save in a serving bowl. Garnish with goji berries and diced Chinese celery. Serve.

龍躉
giant grouper

香檳汁龍躉球

Giant grouper in champagne sauce

材料

龍躉肉 500 克

Ingredients

- 500 g giant grouper fillet

香檳汁

香檳 150 毫升
洋葱 1/2 個（切粒）
紅葱頭 3 湯匙（切粒）
淡忌廉 200 毫升
鹽 1/2 茶匙
菇粉 1 茶匙
熟雞蛋黃 1 隻
橄欖油 1 湯匙
牛油 20 克
雞蛋黃 1 隻

Champagne sauce

- 150 ml champagne
- 1/2 onion (diced)
- 3 tbsp shallot (diced)
- 200 ml whipping cream
- 1/2 tsp salt
- 1 tsp mushroom powder
- 1 yolk of a hard-boiled egg
- 1 tbsp olive oil
- 20 g butter
- 1 raw egg yolk

做法

1. 熟雞蛋黃加橄欖油拌成蛋黃醬，備用。
2. 熱鑊下油，先爆香洋葱及紅葱頭，下香檳，邊攪拌邊煮至略濃稠，下淡忌廉及蛋黃醬，加入菇粉及鹽少許調味，熄火後下生雞蛋黃及牛油拌勻，煮成香檳汁。
3. 龍躉肉切厚片，加少許鹽及蛋白，加入生粉 3 茶匙拌勻，略醃片刻。
4. 燒熱油鑊，下龍躉炸至金黃全熟，盛起上碟，淋上香檳汁即成。

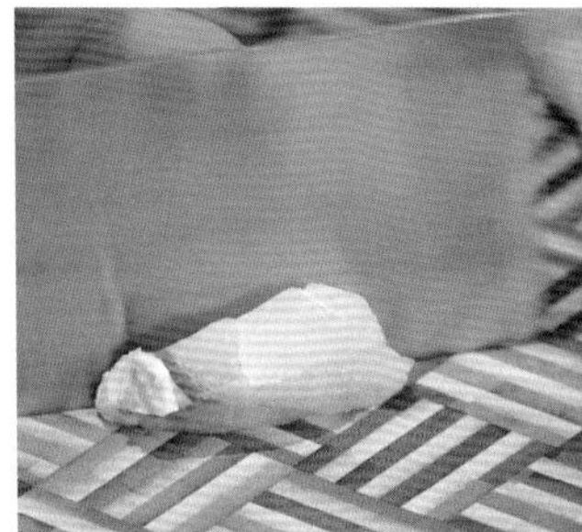

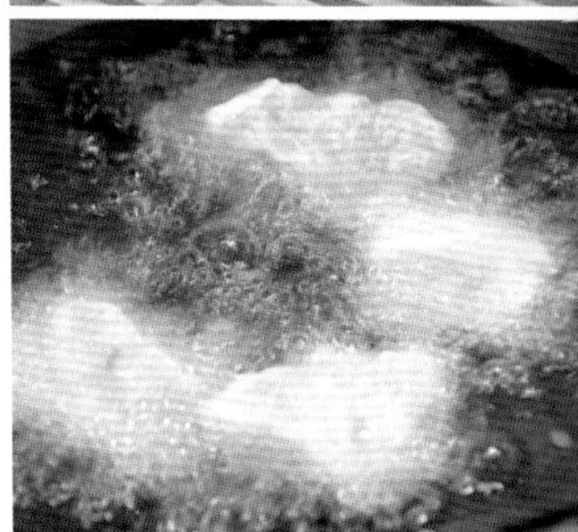

Method

1. Put the yolk of a hard-boiled egg into a bowl. Mash it and add olive oil. Mix well.
2. Heat wok and add oil. Stir-fry onion and shallot. Drizzle with champagne. Cook while stirring until it reduces. Add whipping cream and the egg yolk mixture from step 1. Add mushroom powder and a pinch of salt. Turn off the heat. Stir in raw egg yolk and butter. Mix well. This is the champagne sauce.
3. Slice the grouper thickly. Add a pinch of salt and some egg white. Add 3 tsp of potato starch. Mix well and leave it briefly.
4. Heat wok and add oil. Deep-fry the grouper slices until golden and cooked through. Arrange on a serving plate. Dribble the champagne sauce over. Serve.

必學不敗竅門

- 宜用廚房紙或毛巾印乾龍躉的水分，以免在炸時油分四濺。
- 香檳汁想煮得好，先爆香料頭，再徐徐加入忌廉以小火邊煮邊攪拌，煮至濃稠。
- After slicing the grouper, wipe it dry with a clean towel or paper towel. That would prevent the oil from splattering when deep-frying the grouper.
- To make the best champagne sauce, stir-fry the aromatics until fragrant first. Then slowly pour in the whipping cream while cooking over low heat and stirring continuously until the sauce thickens.

海蝦

marine shrimp

蝦蝦笑 釀秋葵併蝦餅

Stuffed fried okras and minced shrimp patties

材料

- 海蝦 4-6 隻
- 蝦膠 200 克
- 秋葵 10 條
- 熱情果 1 個
- 熱情果醬 1-2 湯匙

Ingredients

- 4 to 6 marine shrimps
- 200 g minced shrimp
- 10 okras
- 1 passionfruit
- 1 to 2 tbsp passionfruit jam

必學不敗竅門

- 建議不要切掉秋葵蒂部，能保留黏液，煎後味道很香，而且不會黏黏的。
- It's advisable to keep the stems on the okras to seal in the slime. The okras taste great after fried and don't feel slimy at all.

COOKING TIP

做法

❶ 秋葵洗淨，瀝乾水分，橫切 1/3 開邊（勿切掉頂部），去籽（籽留用）。
❷ 海蝦去殼、切粒，備用。
❸ 蝦膠及海蝦粒搓勻，撻至起膠，取半份蘸少許生粉，釀入秋葵內。
❹ 剩餘的秋葵及秋葵籽切碎，加入餘下蝦膠搓勻成餅狀。
❺ 熱鑊下油，將秋葵蘸上生粉，蝦膠向下，煎熟至兩面金黃色，上碟。
❻ 蝦餅煎熟至兩面金黃色，上碟。
❼ 熱情果切開挖出果肉，與熱情果醬拌勻成醬汁，淋在秋葵及蝦餅上即成。

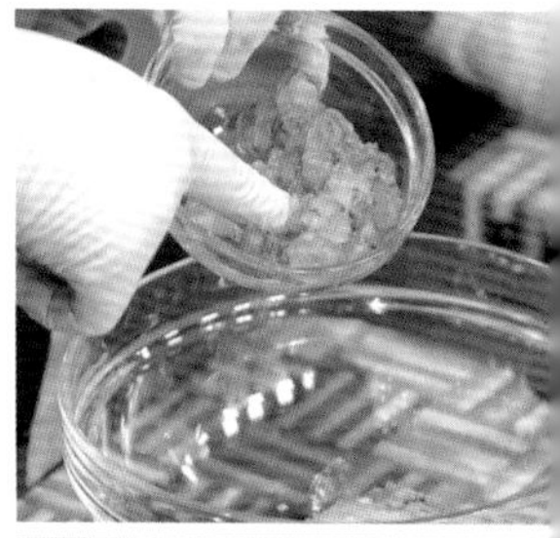

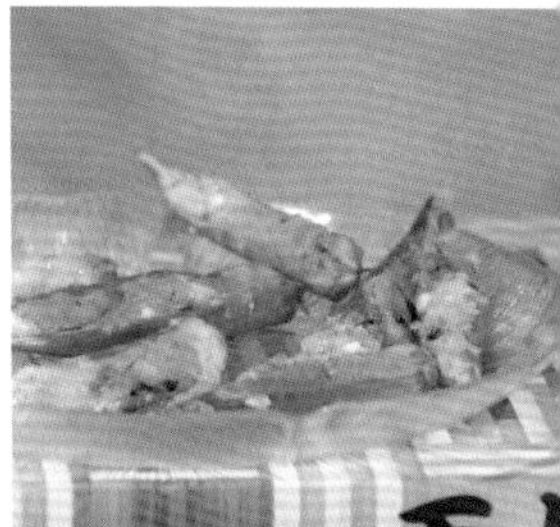

Method

❶ Rinse the okras. Drain well. Make a cut lengthwise to trim off 1/3 of each okra. Do not cut off the stem. Save the trimmed-off parts. Scoop out the seeds and save them for later use.
❷ Shell the shrimps and dice them.
❸ Mix the minced shrimps with diced shrimps. Mix well in a bowl. Lift the mixture off the bowl and slap it back in a few times until sticky. Set aside half of the mixture. Coat the other half in potato starch. Stuff it into the okras.
❹ Finely chop the trimmed-off parts of okras and the seeds from step 1. Add this mixture to the remaining half of the minced shrimp from step 3. Mix well and shape into round patties.
❺ Heat wok and add oil. Coat the stuffed okras in potato starch. Put them into the wok with the minced shrimp side facing down. Fry until both sides golden and cooked through. Save on a serving plate.
❻ Fry the shrimp patties from step 4 in the same wok until both sides golden. Save on a serving plate.
❼ Cut the passionfruit in half. Scoop out the seeds and pulp. Mix with the passionfruit jam. Dribble the mixture over the stuffed okras and shrimp patties. Serve.

花膠
fish maw

鴛鴦墨膠釀花膠

Fish maw rolls with black and white minced cuttlefish

材料

- 花膠 2 隻（已浸發）
- 墨魚膠（白）200 克
- 墨魚膠（黑）200 克
- 蝦膠 80 克
- 紅蘿蔔 6-8 片
- 薑粒 6 湯匙
- 葱 2 條（切幼絲）

蠔油芡

- 蠔油 3 湯匙
- 生粉 1 茶匙
- 水 3 湯匙
- 生抽 1 茶匙
- 麻油 1 茶匙

Ingredients

- 2 fish maws (rehydrated)
- 200 g minced cuttlefish
- 200 g black minced cuttlefish with ink
- 80 g minced shrimp
- 6 to 8 slices carrot
- 6 tbsp diced ginger
- 2 spring onion (finely shredded)

Oyster sauce glaze

- 3 tbsp oyster sauce
- 1 tsp potato starch
- 3 tbsp water
- 1 tsp light soy sauce
- 1 tsp sesame oil

做法

❶ 已發花膠放入沸水內，加少許鹽，蓋好稍焗 5 分鐘待入味，備用。
❷ 雙色墨魚膠各混入半份蝦膠拌勻，再撻數下，備用。
❸ 取花膠抹乾，撲上生粉，分別將黑、白墨魚膠均勻塗滿花膠，從花膠尖的一端向上捲成筒狀，收口處再塗上墨膠封口。
❹ 蒸碟內放上紅蘿蔔片墊底（以防花膠黏着碟），放入花膠筒蒸 10 分鐘，完成後切段，上碟，放入葱絲伴碟。
❺ 薑粒加鹽 1-2 茶匙，放油鑊內爆香，薑粒放在花膠面。
❻ 將蠔油芡材料拌勻，煮滾後淋上花膠面即成。

Method

❶ Put fish maws into a pot of boiling water. Turn off the heat. Season with a pinch of salt. Cover the lid and leave them for 5 minutes. Drain and set aside.
❷ Add 40g of minced shrimp to the white minced cuttlefish. Add the remaining minced shrimp to the black minced cuttlefish with ink. Mix well. Lift each minced cuttlefish off its corresponding bowl and slap it back in a few times.
❸ Wipe dry the fish maws. Coat them in potato starch. Spread the white minced cuttlefish mixture on one fish maw, and the black mixture on the other evenly. Roll each fish maw from the pointy end into a log. Seal the open seam with more minced cuttlefish.
❹ Line a steaming plate with sliced carrot (it would prevent fish maw rolls from sticking to the plate.) Arrange the fish maw rolls on top. Steam for 10 minutes. Slice the fish maw rolls and arrange on a serving plate. Garnish with spring onion on the side.
❺ Add 1 to 2 tsp of salt to the diced ginger. Stir-fry in some oil. Arrange on top of the fish maw rolls.
❻ Mix all ingredients of the oyster sauce glaze well. Pour in the wok and bring to the boil. Dribble over the fish maw rolls. Serve.

必學不敗竅門

- 宜選用較薄身的花膠；先蘸上生粉才釀墨魚膠，捲好後用墨魚膠封口，蒸後的花膠卷才不會鬆散。
- For this recipe, it's advisable to pick thinner fish maws. Coat fish maws in potato starch before spreading the minced cuttlefish over. After rolling the fish maws, seal the open seam with more minced cuttlefish. That would ensure the fish maw rolls won't unroll and turn loose when steamed.

鮑魚

abalone

鮑魚雞粒紅豆荷葉飯

Fried rice with abalone, chicken and red beans in lotus leaf

材料

- 雞扒 1 件
- 罐頭鮑魚 6 隻（300 克）
- 蝦肉 4 隻
- 雞蛋 1 隻
- 瑤柱 40 克
- 薑粒 1 湯匙
- 紅豆 5 湯匙
- 白飯 2-3 碗
- 新鮮及乾荷葉各 1 塊
- 生抽 1 湯匙
- 菇粉 1 茶匙

Ingredients

- 1 boneless chicken thigh
- 6 canned abalones (about 300 g)
- 4 shelled shrimps
- 1 egg
- 40 g dried scallops
- 1 tbsp diced ginger
- 5 tbsp red beans
- 2 to 3 bowls steamed rice
- 1 fresh lotus leaf
- 1 dried lotus leaf
- 1 tbsp light soy sauce
- 1 tsp mushroom powder

做法

1. 乾、鮮荷葉洗淨，以熱水汆燙，印乾水分。乾荷葉放鮮荷葉上面，並塗上一層油備用。
2. 雞扒洗淨、切粒，以少許鹽調味。
3. 蝦肉洗淨、切粗粒；瑤柱洗淨，浸軟，撕成絲。
4. 紅豆洗淨，浸 3 小時以上，隔水蒸半小時至腍身。
5. 罐頭鮑魚取 1 隻留用，其餘切絲。
6. 熱鑊下油，爆香薑粒，下少許鹽炒香，下雞粒、瑤柱絲、鮑魚絲、蝦肉、雞蛋略炒，放入飯翻炒，再加入紅豆炒勻，下適量生抽及菇粉調味。
7. 將原隻鮑魚放荷葉中央，再鋪上炒飯，用荷葉包好，大火隔水蒸 7-8 分鐘，取出上碟，剪開荷葉頂部，即可享用。

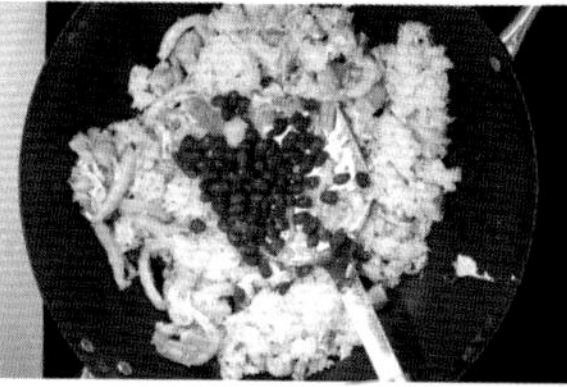

Method

1. Rinse both lotus leaves. Blanch in boiling water. Drain and wipe dry. Lay flat the fresh lotus leaf on the counter. Put the dried one on top. Brush a thin layer of oil on the dried lotus leaf.
2. Rinse the chicken thigh and dice it. Sprinkle with a pinch of salt. Mix well.
3. Rinse the shrimps and dice coarsely. Set aside. Rinse dried scallops and soak them in water till soft. Break into fine shreds.
4. Rinse red beans and soak them in water for at least 3 hours. Drain and steam for 30 minutes until tender.
5. Set aside 1 abalone and keep it in whole. Shred the rest.
6. Heat wok and add oil. Stir-fry ginger until fragrant. Sprinkle with a pinch of salt and toss well. Put in chicken, dried scallops, shredded abalones, shelled shrimps and egg. Toss briefly. Put in the rice. Toss well. Add red beans and toss again. Drizzle with some light soy sauce and sprinkle with mushroom powder.
7. Put the whole abalone at the centre on top of the greased lotus leaf. Top with fried rice. Wrap the lotus leaves into parcel. Steam over high heat for 7 to 8 minutes. Transfer the lotus leaf parcel onto a serving plate. Cut open the top of the parcel. Serve.

必學不敗竅門

- 在乾荷葉面塗油，蒸後的飯粒不會黏着荷葉。
- 宜用比較硬及乾的飯，炒飯才會成功及美味。
- Brushing oil on the dried lotus leaf stops the rice from sticking to it after steamed.
- For fried rice, it's advisable to use day-old rice which is stiffer and dryer in texture. The fried rice would turn out crisper with every grain separable from another.

膏蟹
female mud crab

大蟹蟹膏荷葉飯
Fried rice with crab wrapped in lotus leaf

材料

- 大膏蟹 1 隻
- 新鮮荷葉 1 張
- 乾荷葉 1 張
- 白飯 3-4 碗
- 薑粒 2 湯匙
- 蒜粒 1 湯匙
- 雞蛋 2 隻
- 芹菜粒少許（裝飾）
- 葱花適量（裝飾）
- 芫茜碎適量（裝飾）

炒飯調味料

- 生抽 4 茶匙
- 老抽 2 茶匙
- 鹽 1/4 茶匙
- 糖 1/2 茶匙

Ingredients

示範短片

- 1 female mud crab
- 1 fresh lotus leaf
- 1 dried lotus leaf
- 3 to 4 bowls steamed rice
- 2 tbsp diced ginger
- 1 tbsp diced garlic
- 2 eggs
- diced Chinese celery (as garnish)
- finely chopped spring onion (as garnish)
- finely chopped coriander (as garnish)

Seasoning for fried rice

- 4 tsp light soy sauce
- 2 tsp dark soy sauce
- 1/4 tsp salt
- 1/2 tsp sugar

做法

1. 兩款荷葉洗淨，汆水後印乾水分，放入蒸籠內（先鋪上新鮮的再放乾品），於乾荷葉面塗油，備用。
2. 膏蟹洗淨，掀起蟹蓋，棄去胃、鰓、腸，起蟹鉗，蟹身一開二，備用。
3. 燒熱油鑊，先爆香薑粒，再爆蒜粒，依次放入雞蛋（免拂）及白飯，不停翻炒，期間以鑊鏟按壓白飯至飯粒散開，炒至飯粒粒分明，即可灑入少許鹽，炒勻後熄火，加入生抽、老抽及糖炒勻，放入蒸籠內，再將蟹整齊放上飯面，蟹蓋有膏一面朝上，蓋好蒸籠蓋，大火蒸 8 分鐘。
4. 將蟹蓋反轉放回蟹身上，撒上芹菜粒、葱花及芫茜碎，用荷葉包封好，再蒸 1 分鐘，完成後可在蟹面掃上初榨橄欖油添光澤，趁熱享用。

Method

1. Rinse both lotus leaves. Blanch them in boiling water. Wipe dry. Line a bamboo steamer with the fresh one first. Top with the dried one. Brush cooking oil over the dried lotus leaf.
2. Rinse the crab. Pull off the carapace. Discard the sandy sac, gills and digestive tract. Cut off the pincers. Cut the body into half.
3. Heat wok and add oil. Stir-fry ginger until fragrant. Then put in garlic and toss until fragrant. Pour in the eggs (do not whisk them) and rice in this particular order. Toss continuously until the rice grains are evenly coated in egg. Sprinkle with a pinch of salt and toss well. Turn off the heat and add light soy sauce, dark soy sauce and sugar. Toss again. Transfer the rice into the bamboo steamer over the greased dried lotus leaf. Arrange the crab neatly over the rice. Put in the carapace with the roe side facing up. Cover the steamer and steam over high heat for 8 minutes.
4. Flip the carapace to cover the body of the crab. Sprinkle with Chinese celery, spring onion and coriander. Fold the lotus leaf to wrap well. Steam for 1 more minute. Brush extra-virgin olive oil over the crab to give it a sheen. Serve hot.

必學不敗竅門

- 將兩種荷葉汆水的原因，除了因汆燙後葉較軟易於包裹外，還可以去掉草青味。
- Both fresh and dried lotus leaves should be blanched in boiling water. First off, blanching them make them softer and less resilient, so that you can wrap the rice and filling more easily into a packet. Secondly, blanching helps remove the grassy taste.

OOKING TIPS

龍躉

giant grouper

黃金龍躉夠薑炒飯

Gingery fried rice with giant grouper fillet

必學不敗竅門

- 薑拍扁後切粒，落鑊爆香時只需要略炒，然後讓它繼續煎至香；再加入少許鹽提味，薑味會更濃郁。
- 如果冷飯結成一團，宜用鑊鏟將飯壓鬆。
- Crush the ginger and then dice it. Put it into the hot wok and toss it slightly. Then just let it keep on frying in the wok until fragrant. Adding a pinch of salt helps enhance its aroma.
- If the steamed rice clumps together, just press it with a spatula.

COOKING TIP

材料

- 龍躉肉 300 克
- 連皮薑粒 80 克
- 白飯 3 碗
- 罐頭粟米湯 150 克
- 雞蛋 3-4 隻
- 葱花 10 湯匙
- 上湯適量

Ingredients

- 300 g giant grouper fillet
- 80 g diced ginger (with skin on)
- 3 bowls steamed rice
- 150 g canned cream-style sweet corn
- 3 to 4 eggs
- 10 tbsp finely chopped spring onion
- stock

做法

❶ 龍躉洗淨，切大粒，以適量鹽、胡椒粉調味，加入蛋白約 1 湯匙及生粉 1.5 湯匙撈勻，備用。

❷ 熱鑊下油，下龍躉肉，大火炸至金黃，撈起備用。

❸ 熱鑊下油，先爆香薑粒，加入少許鹽調味，再加入雞蛋 3 隻（免拂），慢火略煎香，下白飯不停翻炒至粒粒分明，加入少許鹽調味，最後放入葱花，灒入適量生抽炒勻，盛起。

❹ 在炒飯的同時，將粟米湯及上湯放熱鑊煮滾，加入雞蛋 1 隻拌勻，最後加少許生粉水埋芡。

❺ 先把龍躉肉排放在炒飯邊，再淋上粟米湯即成。

Method

❶ Rinse the grouper and dice coarsely. Season with salt and ground white pepper. Mix well. Add 1 tbsp of egg white and 1.5 tbsp of potato starch. Mix well.

❷ Heat wok and add oil. Deep-fry the grouper over high heat until golden. Drain.

❸ Heat the same wok and add oil. Stir-fry diced ginger until fragrant. Sprinkle with a pinch of salt and crack in 3 eggs. Do not whisk them. Fry the eggs over low heat until lightly browned. Put in the rice and toss continuously until the rice grains are separable from each other. Season with a pinch of salt and put in spring onion at last. Drizzle with light soy sauce and toss to mix well. Save on a serving plate.

❹ When the rice is frying, pour the cream-style sweet corn and stock into a wok and bring to the boil. Crack in 1 egg and stir well. Stir in potato starch thickening glaze at last.

❺ Arrange the fried grouper next to the fried rice. Dribble the sweet corn sauce over the grouper. Serve.

鮑魚
abalone

紅燒鮑魚粉絲煲

Braised abalones with mung bean vermicelli in clay pot

材料

- 罐頭紅燒鮑魚 10 隻
- 馬尾粉絲 150 克
- 芽菜 80 克
- 脢頭瘦肉 160 克
- 菜脯 1 湯匙（切絲）
- 紅葱頭 6 粒
- 上湯 50-80 毫升
- 鮑汁 2 湯匙
- 老抽 2 湯匙
- 蝦米粉少許
- 鹽、糖少許
- 紅椒絲少許
- 芹菜 1 條（切段）
- 葱 1 條（切段）

Ingredients

- 10 canned braised abalones
- 150 g mung bean vermicelli
- 80 g mung bean sprouts
- 160 g pork shoulder butt
- 1 tbsp salted radish (finely shredded)
- 6 shallots
- 50 to 80 ml stock
- 2 tbsp abalone sauce
- 2 tbsp dark soy sauce
- ground dried shrimps
- salt
- sugar
- red chillies (shredded)
- 1 sprig Chinese celery (cut into short lengths)
- 1 spring onion (cut into short lengths)

做法

❶ 脢頭瘦肉洗淨、切粒，加入少許鹽略醃，備用。

❷ 紅葱頭原粒拍鬆；粉絲用凍水浸軟，瀝乾備用。

❸ 取 1 湯匙罐頭鮑魚內汁與鮑汁混和；老抽加上湯 2 湯匙拌勻，備用。

❹ 燒熱油鑊，放入芽菜快速煸炒至半熟，盛起備用。

❺ 燒熱油鑊，放入紅葱頭爆香，加入脢頭粒快炒，下菜脯絲、鮑魚及粉絲，用筷子邊翻炒邊加入適量上湯，加入鮑魚汁及老抽炒勻，再加入少許糖、蝦米粉調味，最後下紅椒絲、芹菜段、葱段及芽菜，炒至芽菜熟即可放入已燒熱砂鍋內，趁熱享用。

Method

❶ Rinse the pork and dice it. Add a pinch of salt. Mix well and leave it briefly.

❷ Crush the shallots. Set aside. Soak the mung bean vermicelli in cold water until soft. Drain.

❸ Take 1 tbsp of the braising sauce from the canned abalones and add it to the abalone sauce. Mix well and set aside. Add 2 tbsp of stock to the dark soy sauce. Mix well.

❹ Heat wok and add oil. Put in the mung bean sprouts and toss quickly until half cooked. Set aside.

❺ Heat the same wok and add oil. Stir-fry shallots until fragrant. Then put in the pork and toss quickly. Add salted radish, abalones and mung bean vermicelli. Toss with a pair of chopsticks while pouring in the stock. Add abalones sauce and dark soy sauce. Toss well. Season with a pinch of sugar and ground dried shrimps. Lastly put in red chillies, Chinese celery, spring onion and mung bean sprouts. Toss until mung bean sprouts are cooked. Transfer into a heated clay pot. Serve the whole pot.

必學不敗竅門

- 炒粉絲時邊炒邊加進上湯，令粉絲更入味及爽口。
- Pour in the stock while stir-frying the mung bean vermicelli. The mung bean vermicelli would pick up the flavours better this way while keeping its springy texture.

星斑
coral grouper

堂焯麻香星斑

Blanched coral grouper fillet in Sichuan peppercorn soy sauce

材料

- 西星斑 1 條（約 600 克）
- 勝瓜 1/2 條
- 上湯 1 公升
- 葱 2-3 條（切幼絲）

Ingredients

- 1 coral grouper (about 600 g)
- 1/2 angled loofah
- 1 litre stock
- 2 to 3 spring onion (finely shredded)

麻香魚汁

- 新鮮胡椒 2-3 串
- 生抽 5 湯匙
- 魚露 1 茶匙
- 糖 1/2 茶匙
- 花椒油 10 湯匙
- 辣椒油少許
- 紅椒絲少許

Sichuan peppercorn soy sauce

- 2 to 3 sprigs fresh peppercorns
- 5 tbsp light soy sauce
- 1 tsp fish sauce
- 1/2 tsp sugar
- 10 tbsp Sichuan peppercorn oil
- chilli oil
- red chillies (shredded)

做法

1. 勝瓜洗淨、去硬皮，切成圓片，以上湯稍氽燙，整齊排放碟上。
2. 西星斑洗淨、吸乾水分，切開魚頭、魚尾、魚鰭，兩邊魚身起出魚柳，斜切雙飛薄片。魚片放在勝瓜片上；魚頭等部分放另一蒸碟，大火蒸 5 分鐘。
3. 上湯煮滾，加入少許胡椒粉，用湯勺將熱湯重複淋上魚片上，直至魚片完全變白色至焯熟，完成後放入魚頭、魚尾、魚鰭整齊排好。
4. 燒熱鑊，加入少許油爆香新鮮胡椒，加入花椒油、辣椒油及糖燒熱，熄火後加入魚露及生抽拌勻，放入紅椒絲成麻香魚汁。
5. 將麻香魚汁淋上魚片，放上油爆蔥絲伴碟即成。

Method

1. Rinse the angled loofah. Peel off the ridges. Then slice it into discs. Blanch them in stock briefly. Arrange neatly on a plate.
2. Rinse the coral grouper. Wipe dry. Cut off the head, tail and fins. Then fillet the fish and cut into thin butterflied slices at an angle. Arrange the sliced fish over the angled loofah. Set aside. Put the head, tail and fins on a steaming plate. Steam over high heat for 5 minutes.
3. Boil the stock. Add a pinch of ground white pepper. Ladle the hot stock over the sliced fish and drain it off. Repeat this step until the fish is cooked through and turns white. Arrange the steamed head, tail and fins around the sliced fish.
4. To make the Sichuan peppercorn soy sauce, heat wok and add oil. Stir-fry the fresh peppercorns until fragrant. Add Sichuan peppercorn oil, chilli oil and sugar. Bring to the boil. Turn off the heat and add fish sauce and light soy sauce. Mix well.
5. Drizzle the steamed sliced fish with the Sichuan peppercorn soy sauce from step 4. Garnish with finely shredded spring onion that has been fried in oil. Serve.

必學不敗竅門

- 用熱湯慢慢焯熟魚片，可保持魚肉嫩滑。
- Blanching the sliced fish with hot stock slowly to keep the fish velvety and juicy.

OOKING TIPS

瀨尿蝦
mantis shrimp

酒香瀨尿蝦
Wine-marinated mantis shrimps

材料

瀨尿蝦 8 隻
芫茜 1 棵
芹菜粒少許
葱花 2 湯匙
紅椒絲少許

浸汁

紹興酒 300 毫升
糟滷 150 毫升
桂花陳酒 150 毫升
玫瑰露 1 茶匙
糖 1/2 茶匙

Ingredients

- 8 mantis shrimps
- 1 sprig coriander
- diced Chinese celery
- 2 tbsp finely chopped spring onion
- red chillies (shredded)

Wine marinade

- 300 ml Shaoxing wine
- 150 ml distillers grain marinade
- 150 ml aged osmanthus wine
- 1 tsp Chinese rose wine
- 1/2 tsp sugar

必學不敗竅門

- 焯瀨尿蝦的時間不可太久，才能保持肉質嫩滑。
- 瀨尿蝦煮熟後立即放入冰水浸泡，再扭動屈曲外殼，可輕易起肉。
- Do not blanch the mantis shrimps for too long. Otherwise, the flesh would turn rubbery and dry.
- After blanching the mantis shrimps, dunk them into ice water immediately. Twisting the shell makes it easier to remove the shell.

COOKING TIP

做法

❶ 瀨尿蝦買回來用冰水浸泡以防變色，備用。
❷ 芫茜洗淨，浸白開水，加入少許白醋浸泡一會，吸乾水分，切碎備用。
❸ 大碗內放入紹興酒、糟滷、桂花陳酒、玫瑰露及糖拌勻成浸汁，備用。
❹ 瀨尿蝦放沸水內焯熟，放入冰開水內降溫，撈起，吸乾水分。
❺ 將蝦身由上至下屈曲數下，待蝦殼與肉略鬆開，用尖頭剪刀從尾部沿殼邊向頭部剪開，起肉離殼後放回殼內，以保持原貌，備用。
❻ 瀨尿蝦瀝乾水分，浸泡在浸汁內，期間翻動令蝦均勻入味，倒出浸汁，重複浸泡一次，令蝦充分吸收酒香。
❼ 取出蝦，將芫茜碎、芹菜粒及葱花放入浸汁內拌勻，用湯匙輕壓讓香味釋出，再將浸汁淋入蝦身。
❽ 將蝦一對對整齊排入碟內，頭朝上、身微彎，最後淋上適量浸汁，以紅椒絲裝飾即成。

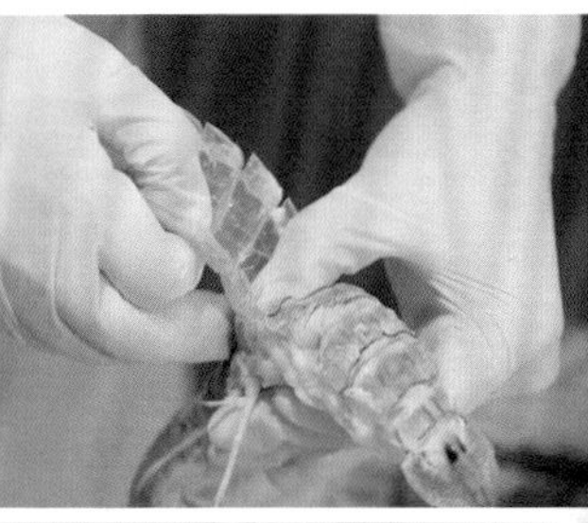

Method

❶ Soak mantis shrimps in ice water to avoid discolouration.
❷ Rinse coriander and soak in cold drinking water with a dash of white vinegar. Drain and wipe dry. Finely chop.
❸ Put wine marinade ingredients into a big mixing bowl. Mix well.
❹ Blanch the mantis shrimps in boiling water until done. Dunk into ice water to cool off instantly. Drain and wipe dry.
❺ Twist the head and tail of a mantis shrimp a few times in opposite directions. This helps separate the meat and the shell. Then cut the point edge of the shell from the tail toward the head with a pair of pointy scissors. Pull the flesh off the shell. Then put it back into the shell. Repeat this step with every shrimp.
❻ Drain the mantis shrimps and soak them in the wine marinade. Flip the shrimps from time to time for them to pick up the seasoning evenly. Then drain the marinade once and pour it back in. That would ensure every shrimp picks up the marinade nicely.
❼ Set the mantis shrimps aside. Put coriander, Chinese celery and spring onion into the wine marinade. Mix well. Press the aromatics with a spoon gently to release the aromas. Then pour the wine marinade onto the shrimps again.
❽ Arrange the mantis shrimps on a serving plate in pairs with the head facing up, and the body curling. Drizzle with some of the wine marinade. Garnish with shredded red chillies. Serve.

蝦
shrimp

香醋脆蝦球
Deep-fried shrimps in aged vinegar sauce

材料

去殼蝦仁 450 克
炸腰果 20 粒（壓碎）
青、紅椒各半個（切角）
炒香芝麻少許

醃料

鹽及胡椒粉各少許
蛋白 1/2 隻
生粉 1 湯匙
麻油少許

香醋汁

陳醋 3 湯匙
鎮江醋 3 湯匙
片糖碎 4-5 茶匙

Ingredients

- 450 g shelled shrimps
- 20 deep-fried cashew nuts (crushed)
- 1/2 green bell pepper (cut into wedges)
- 1/2 red bell pepper (cut into wedges)
- toasted sesames

Marinade

- salt
- ground white pepper
- 1/2 egg white
- 1 tbsp potato starch
- sesame oil

Aged vinegar sauce

- 3 tbsp aged black vinegar
- 3 tbsp Zhenjiang vinegar
- 4 to 5 tsp crushed raw cane sugar slab

做法

1. 蝦仁洗淨，開背、去腸，加入鹽及胡椒粉拌勻，下蛋白、生粉拌勻，再加入麻油略醃備用。
2. 香醋汁材料拌勻，備用。
3. 燒熱炸油，放入青、紅椒略炸，盛起瀝油，備用。
4. 再放入蝦仁，快速炸至 6、7 成熟，盛起備用。
5. 取小鍋，燒熱少許油，放入香醋汁煮熱，加入適量生粉芡煮稠，放入蝦及青、紅椒炒勻，上碟後放入腰果碎伴碟，灑上芝麻即成。

Method

1. Rinse the shrimps. Cut along the back. Devein. Add salt and ground white pepper. Mix well. Add egg white and potato starch. Mix well and drizzle with sesame oil. Leave them briefly.
2. To make the aged vinegar sauce, mix all ingredients until well combined.
3. Heat wok and add oil. Deep-fry the bell peppers briefly. Drain and set aside.
4. Put in the shrimps. Toss quickly until medium-well done. Set aside.
5. In a small pot, heat some oil and put in the aged vinegar sauce. Bring to the boil. Stir in potato starch thickening glaze and cook until it thickens. Put in the shrimps and bell peppers. Toss to coat evenly. Transfer onto a serving dish. Garnish with cashew nuts on the side. Sprinkle with sesames. Serve.

必學不敗竅門

- 醋及糖的份量可依個人口味酌量調整。
- 配搭不同的醋混合使用，再加片糖碎調味，可提升味道。
- 此菜式的蝦球可用煎取代炸。
- You may adjust the amounts of vinegar and sugar according to your personal preference.
- Using different vinegars with raw cane sugar slab adds more depth and complexity to the flavours.
- Instead of deep-frying the shrimps, you may also shallow-fry them.

儲存愈久，陳醋的顏色愈深，存放 1 至 2 年的陳醋品質最上乘。
The colour of aged vinegar turns dark with time. The best ones should have been aged for 1 to 2 years.

OOKING TIPS

雞 / 鮑魚
chicken / abalone

啫啫鮑魚雞煲

Sizzling chicken and abalone in clay pot

材料

- 雞 1 隻
- 罐頭鮑魚 8 隻
- 生筋 8 個
- 薑 3 片（拍扁）
- 紅葱頭 4 粒
- 蒜肉 5 瓣
- 芹菜 1 棵（切段）
- 葱 2 棵（切段）
- 柱侯醬 1 湯匙
- 鮑汁 3 湯匙
- 指天椒少許（切圈）

Ingredients

- 1 chicken
- 8 canned abalones
- 8 deep-fried gluten balls
- 3 slices ginger (crushed)
- 4 shallots
- 5 cloves garlic
- 1 sprig Chinese celery (cut into short lengths)
- 2 spring onion (cut into short lengths)
- 1 tbsp Chu Hau sauce
- 3 tbsp abalone sauce
- bird's eye chilli (cut into rings)

醃料

- 紹興酒 1 湯匙
- 生抽 1 湯匙
- 生粉 1 茶匙
- 鹽少許
- 糖 1/2 茶匙
- 胡椒粉少許

Marinade

- 1 tbsp Shaoxing wine
- 1 tbsp light soy sauce
- 1 tsp potato starch
- salt
- 1/2 tsp sugar
- ground white pepper

做法

❶ 生筋用熱水泡軟，沖掉油分，瀝乾備用。
❷ 雞洗淨、斬件，放入醃料拌勻略醃。
❸ 燒熱油鑊，爆香薑片，下原粒紅葱頭及蒜肉爆香，放入雞件煎香兩面至 6、7 成熟，下柱侯醬爆香，炒至 8 成熟後灒入紹興酒，下罐頭鮑魚、生筋、芹菜段、葱段、鮑汁、指天椒炒勻，放入已燒熱砂鍋內，趁熱享用。

Method

❶ Soak the deep-fried gluten balls in hot water until soft. Rinse off the grease. Drain.
❷ Rinse the chicken and chop into pieces. Add marinade and mix well. Leave it briefly.
❸ Heat wok and add oil. Stir-fry ginger until fragrant. Put in the whole shallots and garlic cloves. Stir-fry till fragrant. Put in the chicken and shallow fry until both sides lightly browned and the chicken is medium-well done. Add Chu Hau sauce and toss until the chicken is almost done. Drizzle with Shaoxing wine. Add abalones, gluten balls, Chinese celery, spring onion, abalone sauce, and bird's eye chilli. Toss to mix well. Transfer into a heated clay pot. Serve hot.

必學不敗竅門

- 雞必須走油或煎至 7 成熟，可保持雞肉嫩滑口感。
- The chicken should be blanched in oil or shallow-fried until almost cooked through, but not quite. That would keep the flesh juicy and tender. Do not overcook it.

OOKING TIPS

鮑魚
abalone

脆皮鮑魚

Battered deep-fried abalones

必學不敗竅門

- 新鮮鮑魚肉與殼緊貼，難以分開，最好的方法是放入攝氏 60-70 度熱水浸數秒，就很輕易除殼。
- 切勿用冷水清洗鮑魚，以免鮑魚肉收緊變韌，建議使用暖水。
- Live abalones attach to their shells very strongly and it's very difficult to shell them when they are alive. The best way to shell them is to soak them in water at 60 to 70°C for a few seconds first before separating the flesh from the shell with a knife.
- Do not rinse abalones in cold water as they tend to shrink and the flesh turns chewy. It's advisable to use warm water instead.

COOKING TIP

材料

- 南非鮮鮑魚 6-8 隻
- 味椒鹽適量
- 七味粉適量
- 西芹絲適量（墊底用）

Ingredients

- 6-8 South African abalones
- peppered salt
- Shichimi tōgarashi (Japanese seven spice mix)
- finely shredded celery (for lining the plate)

炸漿料

- 炸粉 150 克
- 冰水 125 克
- 蛋黃 1 隻
- 油 2 湯匙

Deep-frying batter

- 150 g deep-frying flour
- 125 g ice water
- 1 egg yolk
- 2 tbsp oil

做法

❶ 鮑魚塗上生粉，用牙刷洗淨；放入攝氏 60-70 度熱水內輕浸數秒，去殼取肉，清掉內臟，以暖水洗淨及吸乾水分；先撲上生粉及蘸炸漿，備用。

❷ 燒熱炸油，放入鮑魚炸至微金黃取出，瀝乾油分。

❸ 將已炸的鮑魚回鑊輕炸，盛起瀝油，上碟，放在西芹絲上，撒上味椒鹽及七味粉，趁熱享用。

Method

❶ Rub potato starch on the abalones. Scrub with a toothbrush. Rinse well. Soak in hot water about 60 to 70°C for a few seconds. Shell them and remove the innards. Rinse in warm water and wipe dry. Coat them in potato starch and then dip them in the deep-frying batter.

❷ Heat wok and add oil. Deep-fry the battered abalones until lightly browned. Drain.

❸ Put the abalones back in the wok again and deep-fry briefly. Drain and arrange the abalones on a serving plate lined with finely shredded celery. Sprinkle peppered salt and Shichimi tōgarashi on top. Serve hot.

龍蝦
lobster

香脆欖菜龍蝦乾煎陳村粉

Lobster with pan-fried Chen Cun rice noodles

材料

- 澳洲龍蝦 1 隻
- 陳村粉 500 克（切段）
- 欖菜 2-3 湯匙
- 櫻花蝦 40 克
- 銀芽 40 克
- 紅葱頭 3-4 粒（拍扁）
- 蒜粒 1/2 湯匙
- 指天椒 1 隻（切圈）
- 芹菜少許（切段）
- 葱段少許
- 油葱酥適量

Ingredients

- 1 Australian lobster
- 500 g Chen Cun rice noodles (cut into short lengths)
- 2 to 3 tbsp pickled mustard greens with preserved olives
- 40 g sakura shrimps
- 40 g mung bean sprouts
- 3 to 4 shallots (crushed)
- 1/2 tbsp diced garlic
- 1 bird's eye chilli (cut into rings)
- Chinese celery (cut into short lengths)
- spring onion (cut into short lengths)
- deep-fried shallot bits

醃料

- 鹽少許
- 蛋白 1/2 隻
- 胡椒粉少許
- 生粉適量

Marinade

- salt
- 1/2 egg white
- ground white pepper
- potato starch

做法

❶ 龍蝦切開頭，身直切為二，去掉腸臟，切件，加入醃料拌勻，備用。
❷ 龍蝦頭大火蒸 15-20 分鐘至熟，備用。
❸ 燒熱油鑊，中小火爆香櫻花蝦，瀝油備用。原鑊放入欖菜爆香，瀝油備用。原鑊放入銀芽，大火快炒，瀝油備用。
❹ 燒熱油鑊，放入陳村粉煎香兩面至軟身，灒入生抽 2 茶匙及老抽 1/2 茶匙，放入銀芽炒勻，即可上碟；排入龍蝦頭。
❺ 燒熱炸油鑊，放入龍蝦肉炸至 6、7 成熟，盛起瀝油；再回鑊翻炸，盛起後撒上少許味椒鹽。
❻ 同一時間，燒熱油鑊，爆香紅葱頭，放入蒜粒爆香，下櫻花蝦、欖菜、指天椒、芹菜段、葱段快炒，最後放進炸龍蝦肉炒勻，放上陳村粉面，灑上油葱酥即成。

Method

❶ Cut the head off the lobster. Cut the tail in half along the length. Devein and cut into pieces. Add marinade and mix well.
❷ Steam the lobster head over high heat for 15 to 20 minutes until done.
❸ Heat wok and add oil. Stir-fry sakura shrimps over medium low heat until fragrant. Drain and set aside. In the same wok, stir-fry pickled mustard greens with preserved olives until fragrant. Drain off the oil and set aside. In the same wok, toss mung bean sprouts over high heat quickly. Drain and set aside.
❹ Heat wok and add oil. Fry the Chen Cun rice noodles until golden on both sides and the noodles are softened. Drizzle with 2 tsp of light soy sauce and 1/2 tsp of dark soy sauce. Put in the mung bean sprouts and toss well. Transfer onto a serving dish. Arrange the lobster head over the noodles.
❺ Heat the wok again and add enough oil for deep-frying. Deep-fry the lobster tail pieces until medium-well done. Drain. Then put them back in the wok again and deep-fry for the second time. Drain and sprinkle with peppered salt.
❻ Meanwhile, heat a wok and add oil. Stir-fry shallots until fragrant. Add diced garlic and fry until fragrant. Add sakura shrimps, pickled mustard greens with preserved olives, bird's eye chilli, Chinese celery and spring onion. Toss quickly. Put in the lobster tail pieces. Toss well. Arrange over the bed of Chen Cun rice noodles on the serving dish. Sprinkle with deep-fried shallot bits. Serve.

必學不敗竅門

- 陳村粉買回來後勿冷藏，否則雪硬了難以煎軟。
- 欖菜先爆香，但不能過火，否則會有燶味。
- Do not refrigerate the Chen Cun rice noodles. Just keep them at room temperature. Otherwise, they would turn hard and would be very difficult to soften in the wok.
- Pickled mustard greens with preserved olives need to be fried before used, but make sure you don't heat them too much. Otherwise, they may burn and the dish would taste bitter.

椰皇 / 雪燕

smoked coconut / tragacanth resin

椰皇鮮奶燉雪燕

Double-steamed tragacanth resin and milk in whole smoked coconut

必學不敗竅門

- 椰皇大小不一，份量可按比例調整，蛋白、椰皇水、鮮奶的比例為 2：2：1。
- Smoked coconuts come in different sizes. You may adjust the amounts used according to their sizes. As a rule of thumb, the ratio of egg white: coconut water: milk = 2：2：1.

COOKING TIP

材料

Ingredients

- 椰皇 2 個
- 椰皇水 120 毫升
- 鮮奶 60 毫升
- 雪燕 2 粒（約 10 克）
- 蛋白 4 隻
- 冰糖水 3-5 湯匙

- 2 smoked coconuts
- 120 ml smoked coconut water
- 60 ml milk
- 2 pieces (10 g) tragacanth resin
- 4 egg whites
- 3 to 5 tbsp rock sugar syrup

做法

1. 雪燕早一晚洗淨、浸水，備用。
2. 雪燕取出，倒掉多餘水分，大火蒸熱，按個人口味加入適量冰糖水調味。
3. 椰皇先切去少許底部（讓它蒸時能平穩），破開頂部，倒出椰皇水留用；椰皇以大火蒸 20 分鐘，待用。
4. 蛋白輕拂，倒入椰皇水內拂勻，加入冰糖水後隔篩，與鮮奶拌勻，倒入椰皇內至 8 分滿，封上耐熱保鮮紙蒸 12 分鐘，熄火焗 20 分鐘，最後放入雪燕即成。

Method

1. Rinse and soak tragacanth resin in water overnight. Drain.
2. Drain the tragacanth resin and steam over high heat. Season it with the rock sugar syrup according to your preference.
3. Cut the base of each smoked coconut flat so that they sit securely on flat surface. Cut off the top and drain the coconut water. Save it for later use. Steam the empty smoked coconut over high heat for 20 minutes. Set aside.
4. Whisk the egg whites. Add smoked coconut water. Whisk again. Add rock sugar and mix well. Pass the mixture through wire mesh. Stir in milk. Pour the mixture into the smoked coconuts up to 80% full. Cover with microwave-safe cling film. Steam for 12 minutes. Turn off the heat and let them sit in the steamer for 20 minutes. Arrange the tragacanth resin on top. Serve.

椰青
young coconut

椰子肉柚子沙律
Young coconut pomelo salad

材料

椰青 1 個
牛油生菜 1 個
車厘茄 8 粒（切半）
泰國金柚 1/2 個
炸椰絲 2 湯匙
薄荷葉適量
九層塔葉適量
檸檬葉 3-4 片

Ingredients

- 1 young coconut
- 1 butter lettuce
- 8 cherry tomatoes (halved)
- 1/2 Thai pomelo
- 2 tbsp deep-fried shredded coconut
- mint leaves
- Thai basil leaves
- 3 to 4 Kaffir lime leaves

沙律汁

泰式雞醬 3 湯匙
素魚露 1 湯匙
罐頭椰漿 3 湯匙
濃縮青檸汁 2 湯匙
糖 1 湯匙

Dressing

- 3 tbsp Thai sweet chilli sauce for chicken
- 1 tbsp vegetarian fish sauce
- 3 tbsp canned coconut milk
- 2 tbsp concentrated lime juice
- 1 tbsp sugar

做法

❶ 椰青起肉、切絲。
❷ 檸檬葉洗淨、去梗，切幼絲；柚子肉撕開，備用。
❸ 將所有沙律汁材料放入大碗內拌勻，下柚子肉、車厘茄（保留數粒裝飾）、椰青肉，最後下薄荷葉、九層塔葉、檸檬葉絲、半份炸椰絲拌勻。
❹ 牛油生菜掰開放碟上，倒入椰肉柚子沙律，加上餘下車厘茄、炸椰絲，放上食用花伴碟即成。

Method

❶ Cut off the top of the young coconut. Scoop out the flesh with a metal spoon. Finely shred the coconut flesh.
❷ Rinse the Kaffir lime leaves. Tear off the stems. Finely shred them and set aside. Peel the pomelo and remove the pith. Tear into pieces. Set aside.
❸ In a mixing bowl, put in all the dressing ingredients. Mix well. Put in the pomelo pieces, most of the cherry tomatoes (set aside a few as garnish), and coconut flesh. Lastly add mint leaves, Thai basil leaves, Kaffir lime leaves and half of the deep-fried shredded coconut. Toss well.
❹ Tear leaves off the butter lettuce and arrange on a serving plate. Pour the pomelo salad from step 3 over the lettuce leaves. Arrange the remaining cherry tomatoes and deep-fried shredded coconut on top. Garnish with edible flowers. Serve.

必學不敗竅門

- 薄荷葉輕揉後才放入沙律內，更能帶出香氣。
- 檸檬葉的香氣令沙律帶獨特風味，但份量不要太多，否則掩蓋了其他味道。
- 椰青肉較嫩，適合製成沙律。
- You may roll up the mint leaves and rub them gently before putting them into the salad. That helps release the fragrance and essential oil.
- Kaffir lime leaves give the salad a characteristic fragrance. But do not use too many. Otherwise, their aromas would be too overpowering and cover up other flavours.
- Young coconut flesh is tender and soft. It works better in a salad than regular coconut flesh.

OOKING TIPS

芒果
mango

芒果素蝦米紙卷

Summer rolls with mango and vegetarian shrimps

材料

- 小青瓜 2 條
- 甘筍 1/2 條
- 芒果 1 個
- 素蝦 4 隻
- 五香豆腐乾 1 片
- 越南米紙 4 張
- 羅馬生菜 4 片
- 薄荷葉 12 片
- 羅勒葉 12 片
- 檸檬葉 2 片（切碎）
- 黑芝麻少許
- 胡麻醬適量

Ingredients

- 2 baby cucumbers
- 1/2 carrot
- 1 mango
- 4 vegetarian shrimps
- 1 slice five-spiced dried tofu
- 4 sheets Vietnamese rice paper
- 4 Romaine lettuce leaves
- 12 mint leaves
- 12 sweet basil leaves
- 2 Kaffir lime leaves (finely chopped)
- black sesames
- white sesame salad dressing

醃青瓜料

- 白醋 200 毫升
- 蜂蜜 2-3 湯匙
- 青檸汁 1 湯匙
- 檸檬汁 2 茶匙
- 橙汁 2 湯匙

Pickling brine

- 200 ml white vinegar
- 2 to 3 tbsp honey
- 1 tbsp lime juice
- 2 tsp lemon juice
- 2 tbsp orange juice

做法

❶ 甘筍去皮，與青瓜切長條，備用。
❷ 預備長形深盤，倒入醃青瓜料拌勻，放入青瓜條、甘筍條浸泡 2 小時。
❸ 芒果去皮，切粗條，備用。
❹ 素蝦汆燙後，吸乾水分，橫切一半，切成 3 段，備用。
❺ 豆乾汆燙後，吸乾水分，切條備用。
❻ 米紙沾濕後鋪在濕毛巾上，依次放上羅馬生菜、醃青瓜、甘筍、芒果、豆乾、素蝦，灑上檸檬葉及黑芝麻，放上已輕揉的薄荷葉及羅勒葉，捲起包實，切件上碟，享用時蘸胡麻醬伴吃。

Method

❶ Peel the carrot. Cut carrot and baby cucumbers into long strips.
❷ In a long deep tray, put in all pickling brine ingredients. Mix well and put in the carrot and baby cucumbers. Leave them to soak for 2 hours.
❸ Peel and core the mango. Cut into long strips.
❹ Blanch the vegetarian shrimps in boiling water. Drain and wipe dry. Cut along the length in half. Then cut each half into 3 pieces.
❺ Blanch the dried tofu in boiling water. Drain and wipe dry. Cut into strips.
❻ Dip the rice paper briefly in water and transfer onto a damp towel. Then put on top a romaine lettuce leaf, pickled cucumber and carrot, mango, dried tofu, and vegetarian shrimps. Sprinkle with Kaffir lime leaves and black sesames. Rub the mint leaves and sweet basil leaves between your thumb and index finger. Put them in with the filling. Roll the rice paper firmly. Transfer onto a serving plate and slice it. Serve with the sesame salad dressing on the side.

必學不敗竅門

- 醃青瓜料之酸甜度可因應個人口味而作出調校。
- 越南米紙沾濕後要鋪在濕毛巾上，否則容易乾透難以包裹。
- You may adjust the amount of sugar and white vinegar used in the pickling brine according to your preference.
- After dipping the rice paper in water, you must transfer it onto a damp towel. Otherwise, the rice paper would dry easily and it would be difficult to wrap the ingredients around without breaking.

OOKING TIPS

番茄
tomato

無水素牛肉番茄湯

No-water tomato soup with vegetarian beef

必學不敗竅門

- 此份量較大，可一次做好，存放雪櫃冷藏，食時再翻熱，拌任何肉類或做成火鍋湯底皆可。
- 此湯首要秘訣是熬煮時不停攪拌，令番茄徹底拌勻煮成鮮濃湯底。
- 必須去掉番茄皮，皮容易黏着鍋底，也令番茄難以煮溶。
- For the amounts of ingredients listed here, you can make a big batch of soup. Just keep the leftover in the freezer and reheat it before serving. You may stir in any meat or use it as a hot pot soup base.
- The key to this recipe is to continuously stir the soup. That helps breaking down the tomatoes into a dense broth.
- You must peel the tomatoes before using. Otherwise, the skin would stick to the pot and the tomatoes would take more time to break down.

COOKING TIP

材料

- 番茄 40 個
- 素牛肉 400 克
- 初榨橄欖油少許
- 羅勒葉 2 片（裝飾）
- 鹽適量

Ingredients

- 40 tomatoes
- 400 g vegetarian beef
- extra-virgin olive oil
- 2 sweet basil leaves (as garnish)
- salt to taste

做法

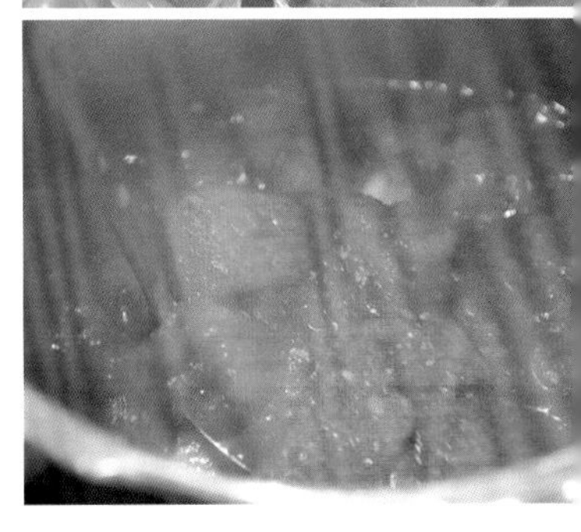

1. 素牛肉加入少許鹽及胡椒粉拌勻，備用。
2. 番茄洗淨，用小刀劃「十」字，放入沸水稍汆燙，取出浸冰水，逐一去掉番茄皮，每個切成 8 件，備用。
3. 燒熱湯鍋，分批放入番茄，逐少加入適量鹽調味，邊煮邊不停攪拌，熬煮 20-30 分鐘，直至番茄開始煮爛、番茄汁充分釋出成番茄湯底，調校至小火。
4. 同一時間，燒熱另一油鑊，放入素牛肉煎熟，加進番茄湯底內拌勻，盛起，放入初榨橄欖油及羅勒葉即成（可按個人口味加入少許糖調味）。

Method

1. Add a pinch of salt and ground white pepper to the vegetarian beef. Mix well.
2. Rinse the tomatoes. Cut a “X” on the bottom of each tomato. Blanch in boiling water briefly. Drain and dunk into ice water. Peel all tomatoes and cut each into eighths.
3. Heat a soup pot. Put in the tomatoes in batches. Season with a little salt each time and taste it. Cook while stirring continuously for about 20 to 30 minutes until the tomatoes break down and turn into a thick broth. Turn to low heat.
4. Meanwhile, heat another wok. Add some oil and fry the vegetarian beef until cooked through. Add the vegetarian beef to the tomato soup. Mix well. Pour the soup into serving bowls. Add a dash of extra-virgin olive oil and a sweet basil leaf. Serve.

素牛肉
vegetarian beef

素西湖牛肉羹
Vegetarian West Lake beef soup

材料

- 素牛肉 450 克
- 盒裝煎煮豆腐 1 盒
- 鮮冬菇 6 朵
- 冬菇水 1 公斤
- 芫茜 2 棵（切碎）
- 鹽 1/2 茶匙
- 糖 1/4 茶匙
- 生粉 1-2 湯匙（用水調勻）
- 胡椒粉少許

Ingredients

- 450 g vegetarian beef
- 1 pack firm tofu for pan-frying
- 6 fresh shiitake mushrooms
- 1 kg water used for soaking dried shiitake mushrooms
- 2 sprigs coriander (finely chopped)
- 1/2 tsp salt
- 1/4 tsp sugar
- 1 to 2 tbsp potato starch (mixed with water into a thickening glaze)
- ground white pepper

做法

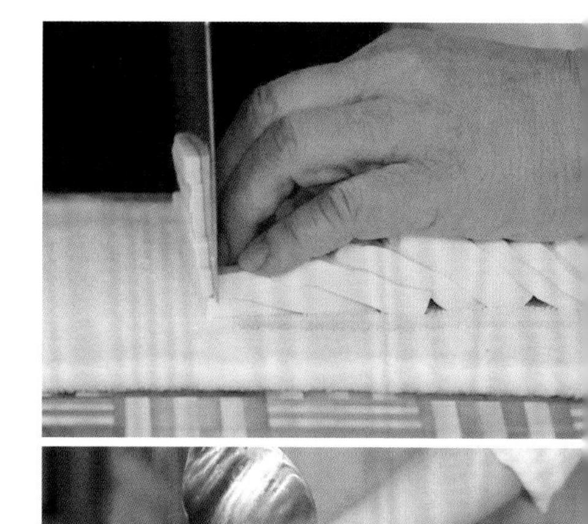

1. 豆腐切幼條；芫茜洗淨切碎；鮮冬菇洗淨，去蒂，切片，菇蒂留用。
2. 冬菇水放入鍋內，加入油、菇蒂及適量開水煮滾。
3. 熱鑊下油，下素牛肉煎香，加進冬菇水內，用篩隔走油及泡沫，依次下冬菇片、豆腐幼條煮滾，下鹽、糖調味，加入芫茜碎，逐少加入生粉水埋芡，最後撒少許胡椒粉，盛起即成。

Method

1. Cut tofu into fine strips. Set aside. Rinse and chop coriander finely. Set aside. Rinse the shiitake mushrooms and slice them. Set aside the stems for later use.
2. Pour the soaking water from dried shiitake mushrooms into a pot. Add oil, stems of shiitake mushrooms and some drinking water. Bring to the boil.
3. Heat wok and add oil. Shallow-fry the vegetarian beef until fragrant. Put the vegetarian beef into the soup from step 2. Skim off the oil and foam on top with a mesh ladle. Then put in shredded shiitake mushrooms and tofu in that particular order. Season with salt and sugar. Add coriander. Slowly stir in potato starch thickening glaze while stirring continuously. Sprinkled with a pinch of ground white pepper. Serve.

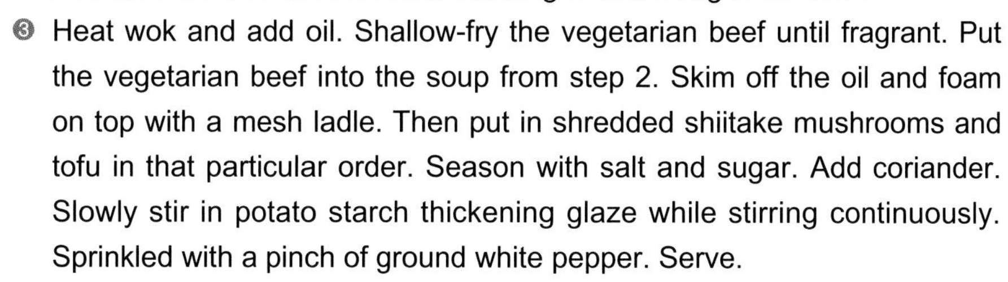

必學不敗竅門

- 素牛肉加入冬菇水後，必須用隔篩濾去泡沫，令湯羹更清澈。
- After adding the vegetarian beef to the soup, make sure you skim off the foam with a mesh ladle. That would keep the soup crystal clear.

OOKING TIPS

素水煮魚

Vegetarian Sichuan boiled fish

材料

- 白靈菇 1 個
- 黃耳 60 克
- 榆耳 60 克
- 木耳 1 湯匙
- 大豆芽 150 克
- 娃娃菜 3 棵
- 硬豆腐 1 件
- 枝竹 2-3 條
- 白蘿蔔 1 條
- 薯粉條 150 克
- 指天椒 2 隻（切圈）
- 芫茜 3 棵（切段）
- 麻辣醬 1-2 茶匙
- 素高湯 300 毫升
- 乾青花椒 1-2 茶匙
- 乾紅花椒 1-2 茶匙

Ingredients

- 1 pleurotus nebrodensis mushroom
- 60 g yellow ear fungus
- 60 g elm ear fungus
- 1 tbsp wood ear fungus
- 150 g soybean sprouts
- 3 baby Napa cabbages
- 1 cube firm tofu
- 2 to 3 dried beancurd sticks
- 1 radish
- 150 g sweet potato starch noodles
- 2 bird's eye chillies (cut into rings)
- 3 sprigs coriander (cut into short lengths)
- 1 to 2 tsp Mala sauce
- 300 ml vegetarian stock
- 1 to 2 tsp dried green Sichuan peppercorns
- 1 to 2 tsp dried red Sichuan peppercorns

麻辣油

- 花椒油 100 毫升
- 乾紅椒 15-20 隻
- 紅椒油 100 毫升
- 麻辣醬 5-6 湯匙

Mala oil

- 100 ml Sichuan peppercorn oil
- 15 to 20 dried red chillies
- 100 ml red chilli oil
- 5 to 6 tbsp Mala sauce

做法

1. 白靈菇抹乾淨，切片，備用。
2. 黃耳洗淨、浸軟，掰開成小粒；榆耳沖洗、浸軟後再洗淨；木耳洗淨、浸軟。
3. 薯粉條洗淨，以暖水浸 25 至 30 分鐘，備用。
4. 硬豆腐切小粒；娃娃菜洗淨、切段；枝竹洗淨、浸軟；白蘿蔔洗淨，去皮切件，備用。
5. 熱鑊下油，爆香指天椒、乾青、紅花椒，加進麻辣醬略爆炒，下素高湯及同等份量水煮滾，下少許鹽、糖調味。
6. 依次放進白蘿蔔、娃娃菜、黃耳、榆耳、大豆芽、枝竹、白靈菇，蓋好煮約 15 分鐘，下薯粉條、木耳、豆腐粒煮開。
7. 另一熱鑊下花椒油、乾紅椒，再下紅椒油及麻辣醬，煮滾備用。
8. 芫茜放入鍋內，淋上麻辣油即成。

Method

1. Wipe down the pleurotus nebrodensis mushroom. Slice it.
2. Rinse the yellow ear fungus. Soak it in water till soft. Tear into small pieces. Set aside. Rinse and soak the elm ear fungus in water till soft. Drain and set aside. Rinse the wood ear fungus and soak it in water till soft. Drain.
3. Rinse the sweet potato starch noodles. Soak in warm water for 25 to 30 minutes. Drain.
4. Dice the firm tofu finely. Set aside. Rinse the baby Napa cabbages. Cut into short lengths. Set aside. Rinse and soak beancurd sticks in water till soft. Drain and set aside. Rinse the radish. Peel and cut into pieces.
5. Heat wok and add oil. Stir-fry bird's eye chillies, dried green and red Sichuan peppercorns until fragrant. Add Mala sauce and toss briefly. Add vegetarian stock and the same volume of water. Bring to the boil. Season with salt and sugar.
6. Then put in the following ingredients in this particular order: radish, baby Napa cabbages, yellow ear fungus, elm ear fungus, soybean sprouts, beancurd sticks, and pleurotus nebrodensis mushroom. Cover the lid and cook for 15 minutes. Put in the sweet potato starch noodles, wood ear fungus and tofu. Bring to the boil.
7. Meanwhile make the Mala oil in another wok. Heat wok and put in the Sichuan peppercorn oil and dried red chillies. Toss well. Add red chilli oil and Mala sauce. Bring to the boil.
8. Add coriander to the wok from step 6. Drizzle with the Mala oil from step 7. Serve.

必學不敗竅門

↑ 市面上較少見的白靈菇
Pleurotus nebrodensis mushroom.

- 榆耳容易藏有沙粒，要徹底清理及洗淨。
- 用硬豆腐除了容易切成塊外，也不容易煮爛。
- Elm ear fungus tends to carry sand. Make sure you clean it well.
- This recipe calls for firm tofu because it's easier to dice and it won't break down as easily after prolonged cooking.

COOKING TIPS

素魚

vegetarian fish

素酸菜魚

Vegetarian fish with pickled mustard greens

示範短片

做法

❶ 鹹酸菜加少許鹽，用水浸泡 20-30 分鐘，沖洗後擠乾，切段備用。
❷ 海帶浸泡至鹹味減退（不同產地浸泡時間有別），切條備用。
❸ 雞髀菇洗淨，用手撕開一片片；豆腐切成 5 件；素魚每件切成 4 片，備用。
❹ 以白鑊將大豆芽及鹹酸菜分別烘乾，盛起備用。
❺ 燒熱油鑊，放入素魚煎至兩面金黃，盛起備用。
❻ 燒熱油鑊，先爆香薑粒，放入蒜粒及鹹酸菜爆炒，再下指天椒及花椒粒炒香，加入沸水約 1 公升，放入芫茜根煮至香味釋出後取走，放入大豆芽、雞髀菇、金菇、海帶、素魚同煮至滾。
❼ 取另一平底鍋，放入豆腐排好燒熱，放入步驟⑥的所有材料。
❽ 同步燒熱小鍋，放入花椒油煮滾，加入芫茜爆香，下麻油，倒入⑦平底鍋，最後下素魚露調味，撒上炒香芝麻即成。

必學不敗竅門

- 大豆芽記緊以白鑊烘乾才煮，能夠去掉豆腥味。
- 若鹹酸菜太鹹，建議用鹽水浸泡，能將鹹味有效地帶走。
- Soybean sprouts need to be fried in a dry wok until dry before used. This step would remove the grassy taste of the soy bean sprouts.
- If the pickled mustard greens are too salty, you may soak them in salted water to remove the saltiness.

COOKING TIP

材料	Ingredients
素白帶魚 3 件	3 pieces vegetarian hairtail
雞髀菇 2 條	2 king oyster mushrooms
金菇 200 克	200 g enokitake mushrooms
鹹酸菜 1 棵	1 head pickled mustard greens
海帶 80 克	80 g dried kelp
大豆芽 100 克	100 g soybean sprouts
布包豆腐 2 件	2 cubes cloth-wrapped tofu
薑粒適量	diced ginger
蒜粒適量	diced garlic
指天椒 5 隻（去蒂）	5 bird's eye chillies (stems cut off)
花椒 1 湯匙	1 tbsp Sichuan peppercorns
芫茜根 5 棵	5 stems coriander roots
素魚露 1 湯匙	1 tbsp vegetarian fish sauce
炒香芝麻適量	toasted sesames

芫茜花椒油	Coriander Sichuan peppercorn oil
花椒油 150 毫升	150 ml Sichuan peppercorn oil
芫茜 5 棵（切碎）	5 sprigs coriander (finely chopped)
麻油 1 湯匙	1 tbsp sesame oil

Method

1. Add a pinch of salt to the pickled mustard greens. Soak in water for 20 to 30 minutes. Rinse and squeeze dry. Cut into short lengths.
2. Soak kelp in water to make it less salty (the soaking time depends on where the kelp comes from). Cut into strips.
3. Rinse the king oyster mushrooms. Tear into slices. Set aside. Cut tofu into 5 pieces. Set aside. Cut each piece of vegetarian hairtail into 4 slices.
4. Heat a dry wok. Put in soybean sprouts and pickled mustard greens. Fry until dry. Set aside.
5. Heat wok and add oil. Fry the vegetarian hairtail until both sides golden. Set aside.
6. Heat the same wok and add oil. Stir-fry diced ginger until fragrant. Put in diced garlic and pickled mustard greens. Toss well. Add bird's eye chillies and Sichuan peppercorns. Toss until fragrant. Add 1 litre of boiling water. Put in the coriander roots until their flavour is infused. Remove the coriander roots. Put in soybean sprouts, king oyster mushrooms, enokitake mushrooms, kelp and vegetarian hairtail. Bring to the boil.
7. In a pan, arrange the tofu neatly and heat. Pour the mixture from step 6 over the tofu.
8. Meanwhile, heat a small pot and put in the Sichuan peppercorn oil. Cook until smoking hot. Put in the coriander and add sesame oil. Pour the mixture over the ingredients in the pan. Lastly season with vegetarian fish sauce. Sprinkle with toasted sesames. Serve.

素雞
vegetarian chicken

麻辣毛豆爆素雞

Fried vegetarian chicken with soybeans in Mala sauce

材料

- 毛豆 200 克
- 素雞漢堡扒 4 件
- 雪菜 60 克（老）
- 紅椒絲少量
- 薑粒 1 湯匙
- 花椒油 1 湯匙
- 辣椒油半湯匙

Ingredients

- 200 g young soybeans
- 4 vegetarian chicken hamburger steaks
- 60 g aged Xue Cai (salted potherb mustard)
- shredded red chillies
- 1 tbsp diced ginger
- 1 tbsp Sichuan peppercorn oil
- 1/2 tbsp chilli oil

調味料

- 鹽及糖各少許
- 紹興酒適量

Seasoning

- salt
- sugar
- Shaoxing wine

做法

1. 雪菜洗淨、浸水，擠乾水分，切小段，以白鑊炒香。
2. 毛豆洗淨，去殼取仁，焯至半熟，備用。
3. 素雞漢堡扒切成粗條，備用。
4. 熱鑊下油，爆香薑粒，下素雞扒、毛豆仁炒至水分收乾，下雪菜、紅椒絲炒香，加入糖、鹽調味，灒酒，最後下花椒油、辣椒油兜勻，上碟即成。

Method

1. Rinse Xue Cai and soak it in water. Drain and squeeze dry. Cut into short lengths. Fry in a dry wok until fragrant.
2. Rinse young soybeans. Discard the pods and use the kernels only. Blanch in boiling water until half-cooked. Drain.
3. Cut the vegetarian chicken steak into thick strips.
4. Heat wok and add oil. Stir-fry diced ginger until fragrant. Put in the vegetarian chicken steaks and young soybeans. Toss until the liquid reduces. Add Xue Cai and shredded red chillies. Season with salt and sugar. Drizzle with Shaoxing wine. Lastly sprinkle with Sichuan peppercorn oil and chilli oil. Toss well and serve.

必學不敗竅門

- 雪菜下鑊前要洗淨、泡水，擠乾後以白鑊烘透，可去除過多鹹味及青草味。
- Xue Cai needs to be rinsed and soaked in water first. Then squeeze dry and fry in a dry wok until no more liquid oozes out. This step helps make it less salty and remove the grassy taste.

脆炸金菇豆腐併紫蘇夾芝士

Deep-fried tofu with enokitake mushrooms and Shiso leaf tempura

材料

- 金菇 80 克
- 甘筍 1/4 條
- 盒裝煎炸豆腐 1 盒
- 芝士件 3 件
- 新鮮紫蘇葉 6 片
- 味椒鹽適量
- 雪菜少許（新）

Ingredients

- 80 g enokitake mushrooms
- 1/4 carrot
- 1 pack firm tofu for pan-frying
- 3 pieces cheddar cheese
- 6 fresh Shiso leaves
- peppered salt
- young Xue Cai (salted potherb mustard)

天婦羅炸漿

- 天婦羅粉 80 克
- 冰水適量
- 蛋黃 1 隻
- 油少許

Tempura batter

- 80 g tempura flour mix
- ice water
- 1 egg yolk
- oil

脆豆腐炸粉

- 麵粉 170 克
- 生粉 120 克
- 粘米粉 40 克
- 粟粉 75 克
- 梳打粉 18 克
- 泡打粉 2 克
- 鹽 10 克

Flour crust for tofu

- 170 g flour
- 120 g potato starch
- 40 g long-grain rice flour
- 75 g cornstarch
- 18 g baking soda
- 2 g baking powder
- 10 g salt

做法

1. 準備天婦羅炸漿：天婦羅粉加冰水調稀，下蛋黃及油拌勻。
2. 將脆豆腐炸粉所有材料拌勻。
3. 甘筍洗淨，去皮、切絲；金菇洗淨，吸乾水分；紫蘇葉洗淨，吸乾水分；雪菜洗淨、浸鹽水，擠乾水分，取葉備用。
4. 豆腐切一口大小，撲上炸粉。燒熱油，炸籬先沾上熱油，放入豆腐粒，再放進熱油炸至金黃，上碟後撒上味椒鹽。
5. 金菇、甘筍絲放大碗內，撲上少許生粉，均勻地蘸上天婦羅炸漿，放熱油炸至金黃，上碟。
6. 芝士件撲上少許生粉，蘸上少許天婦羅炸漿，取兩片紫蘇葉夾着芝士輕壓實，再均勻蘸上天婦羅炸漿，放熱油炸至金黃，上碟。
7. 雪菜直接放熱油炸脆，上碟放金菇甘筍絲上即成。

Method

1. To make the tempura batter, add ice water to the tempura flour mix. Stir into a thin paste. Add egg yolk and oil. Mix well.
2. To make the flour crust for tofu, mix all ingredients together.
3. Rinse and peel carrot. Shred it. Set aside. Rinse the enokitake mushrooms and wipe dry. Set aside. Rinse the Shiso leaves. Wipe dry and set aside. Rinse the Xue Cai. Soak in salted water and drain. Squeeze dry. Use the leaves only. Discard the stems.
4. Cut tofu into bite-sized cubes. Coat them each in the flour crust mixture. Heat oil in a wok and dip the strainer ladle into the oil first. Put the diced tofu into the strainer ladle and deep-fry until golden. Drain and transfer onto a plate. Sprinkle with peppered salt.
5. Put enokitake mushrooms and carrot into a mixing bowl. Coat them in potato starch. Then coat them evenly in the tempura batter. Deep-fry in hot oil until golden. Drain and save on a serving plate.
6. Coat each slice of cheese in potato starch, then dip in the tempura batter. Sandwich the cheese in between the Shiso leaves. Dip into tempura batter again and deep-fry in hot oil until golden. Drain and save on a serving plate.
7. Deep-fry the Xue Cai directly in hot oil until crispy. Drain and arrange over the enokitake mushroom tempura. Serve.

必學不敗竅門

- 芝士下油鑊炸前，必須放在雪櫃冷藏至硬，以免芝士快速溶掉。
- Before using, make sure you keep the cheese in the fridge so that it won't melt too quickly when fried.

蘆筍
asparagus

蘆筍勝瓜紫菜腐皮卷

Beancurd skin rolls with asparagus, angled loofah and laver

材料

- 蘆筍 3 條
- 勝瓜 1/3 條
- 紫菜 1 張（圓形）
- 鮮腐皮 2 張
- 素上湯 600 毫升
- 薑蓉 2 湯匙
- 鹽少許

Ingredients

- 3 asparaguses
- 1/3 angled loofah
- 1 sheet laver seaweed (round)
- 2 sheets fresh beancurd skin
- 600 ml vegetarian stock
- 2 tbsp grated ginger
- salt

素蠔油汁

- 素蠔油 2 湯匙
- 水 2 湯匙
- 糖 1/4 茶匙
- 生抽 1 茶匙
- 生粉 1 茶匙

Glaze

- 2 tbsp vegetarian oyster sauce
- 2 tbsp water
- 1/4 tsp sugar
- 1 tsp light soy sauce
- 1 tsp potato starch

做法

1. 紫菜洗淨、浸軟，抹乾水分，加適量麻油、胡椒粉及鹽略醃。
2. 勝瓜洗淨、去硬皮，開半及去瓤，切段（約 5 厘米），備用。
3. 蘆筍洗淨，削掉外皮粗纖維部分，放入素上湯內汆燙入味，盛起，用冰水略浸，抹乾水分，切段（與勝瓜相若）備用。
4. 腐皮洗淨，切成 5 厘米寬，放上蘆筍、勝瓜，捲成腐皮卷，排在蒸碟，取少量紫菜放在腐皮卷上。
5. 熱鑊下油，爆香薑蓉，下少許鹽調味，放在腐皮卷上，隔水蒸 3 分鐘。
6. 素蠔油汁材料調勻，加熱煮成素蠔油汁，淋上腐皮卷上即成。

Method

1. Rinse laver seaweed. Soak in water till soft. Drain and wipe dry. Add sesame oil, ground white pepper and a pinch of salt. Mix well and leave it briefly.
2. Rinse and peel the angled loofah. Cut in half and remove the seeds. Cut into pieces about 5 cm long. Set aside.

3. Rinse the asparaguses. Peel off the tough fibrous skin. Blanch in boiling vegetarian stock until flavourful. Drain. Soak in ice water briefly. Wipe dry and cut into short lengths about the same size as the angled loofah.
4. Rinse the beancurd skin. Cut into pieces about 5 cm wide. Arrange asparagus and angled loofah on top. Roll into a rectangular packet. Arrange on a steaming plate. Put some laver on each beancurd skin roll.
5. Heat wok and add oil. Stir-fry grated ginger and season with a pinch of salt. Arrange ginger over the beancurd skin rolls. Steam the rolls for 3 minutes.
6. Mix all glaze ingredients until well combined. Transfer into a pot and heat it up. Dribble over the steamed beancurd skin rolls. Serve.

必學不敗竅門

- 蘆筍先用上湯煨至入味，令菜式更添美味，再浸冰水能保持翠綠色澤。
- Blanch the asparaguses in stock until flavourful first. That would give the asparagus more flavours to begin with, so that the beancurd skin rolls would taste better. Blanching the asparaguses in ice water helps keeping them bright green.

OOKING TIPS

猴頭菇
monkey head mushroom

香橙猴頭菇

Dried-fried monkey head mushrooms in orange glaze

材料

素猴頭菇丁 400 克（已調味）
生粉 3 湯匙

Ingredients

- 400 g diced monkey head mushrooms (seasoned)
- 3 tbsp potato starch

濃縮橙醬

濃縮橙汁 420 毫升
橙 1 個
吉士粉 1 茶匙
白醋 380 毫升
水 75 毫升
糖 100 克
鹽 1/4 茶匙

Orange glaze

- 420 ml concentrated orange juice
- 1 orange
- 1 tsp custard powder
- 380 ml white vinegar
- 75 ml water
- 100 g sugar
- 1/4 tsp salt

做法

❶ 素猴頭菇丁解凍，加入生粉拌勻，備用。
❷ 橙洗淨，起肉、切粒，備用。
❸ 吉士粉加 3.5 茶匙水調成薄芡，備用。
❹ 熱鑊下油，下猴頭菇丁半煎炸至金黃，盛起，隔油備用。
❺ 將所有濃縮橙醬材料（鹽及吉士粉除外）拌勻，放進熱鑊煮滾，加鹽調味，盛起，取約 100 毫升，加入吉士粉芡煮至濃稠。
❻ 另一熱鍋下猴頭菇丁炒熱，加入濃縮橙醬拌勻，上碟後以食用花點綴即成。

Method

❶ Thaw the diced monkey head mushrooms. Add potato starch and mix well.
❷ Rinse the orange. Peel and tear into segments. Remove the pith. Dice the flesh.
❸ Add 3.5 tsp of water to custard powder. Mix well.
❹ Heat wok and add oil. Cook the monkey head mushrooms in semi-deep frying manner until golden. Drain and set aside.
❺ Mix all glaze ingredients (except salt and custard powder) until well combined. Pour into a hot wok and bring to the boil. Season with salt. Set aside. Pour 100 ml of the orange juice mixture into another pot. Stir in the custard powder mixture from step 3. Cook while stirring until it thickens.
❻ In another wok, stir-fry the monkey head mushrooms until heated through. Pour in the orange glaze. Toss well. Save on a serving plate. Garnish with edible flowers. Serve.

必學不敗竅門

- 煮醬汁時要不斷攪拌；而且下少許鹽調味，能令酸甜味更突出。
- 加入吉士粉芡，令橙醬的顏色更美觀。
- When you make the orange glaze, make sure you keep stirring it while heating. Seasoning with a pinch of salt would accentuate the sourness.
- Adding custard powder mixture to the glaze makes the orange colour brighter so that the dish looks more appetizing.

OOKING TIPS

土多啤梨 / 南瓜
strawberry / pumpkin

土多啤梨炸南瓜

Deep-fried pumpkin in strawberry glaze

材料

- 士多啤梨 6 粒
- 紅桑子 6-8 粒
- 南瓜 1/2 個
- 黃心番薯 1 個
- 素雞 2 條
- 生粉 4-5 湯匙
- 士多啤梨果醬 2-3 湯匙

Ingredients

- 6 strawberries
- 6 to 8 raspberries
- 1/2 pumpkin
- 1 yellow sweet potato
- 2 strips vegetarian chicken
- 4 to 5 tbsp potato starch
- 2 to 3 tbsp strawberry jam

炸漿

- 炸粉 200 克
- 冰水

Deep-frying batter

- 200 g deep-frying batter mix
- ice water

做法

1. 紅桑子、士多啤梨分別洗淨，瀝乾水分。士多啤梨去葉片，每粒切成 8 份，備用。
2. 素雞以滾刀切件，備用。
3. 番薯及南瓜去皮，以滾刀切件，放入沸水煮約 1 分鐘，盛起，瀝乾水分，備用。
4. 炸粉及冰水拌勻（比例約炸粉 1：冰水 0.8，按濃稠度調較），加入少許油，拌勻成炸漿。
5. 番薯、南瓜、素雞蘸上生粉，再沾上炸漿，下油鑊炸熟至金黃色，盛起。
6. 熱鑊下油，下士多啤梨粒、紅桑子以中小火煮開，加入士多啤梨果醬及少許清水，煮至濃稠，下番薯、南瓜、素雞拌勻，上碟，以士多啤梨片裝飾即成。

Method

1. Rinse raspberries and strawberries separately. Drain. Remove the leaves and stems on the strawberries. Cut each into eighths.
2. Cut the vegetarian chicken into random wedges while rolling it on a chopping board.
3. Peel sweet potato and pumpkin. Cut into random wedges. Blanch in boiling water for about 1 minute. Drain well.
4. Mix the deep-frying batter mix with ice water (about 1 part batter mix to 0.8 part ice water, adjust the amounts of water added according to the consistency.) Add a dash of oil. Stir into a smooth batter.
5. Coat sweet potato, pumpkin and vegetarian chicken in potato starch. Then dip them into the deep-frying batter. Heat a wok of oil and deep-fry them until golden. Drain.
6. Heat wok and add oil. Put in the diced strawberries and raspberries. Cook over medium-low heat. Add strawberry jam and some water. Cook until well mixed and thick. Put in the fried sweet potato, pumpkin and vegetarian chicken. Toss to coat evenly. Save on a serving plate. Garnish with sliced strawberries. Serve.

必學不敗竅門

- 炸粉用冰水調開，再加入少許油，炸出來的食物更香口、更鬆脆。
- Mix the deep-frying batter mix with ice water. Then add a dash of oil. That would make sure the crust would be fluffy and light after deep-fried.

雞髀菇

king oyster mushroom

三杯雞髀菇

Three-cup king oyster mushrooms

材料	Ingredients
雞髀菇 3 隻	• 3 king oyster mushrooms
薑 6 片	• 6 slices ginger
蒜粒 1 湯匙	• 1 tbsp diced garlic
九層塔 3 棵（取葉）	• 3 sprigs Thai basil (leaves only)
老抽 1 湯匙	• 1 tbsp dark soy sauce
芝麻油 2 湯匙	• 2 tbsp sesame oil
黑麻油 1 湯匙	• 1 tbsp black sesame oil
紹興酒 2 湯匙	• 2 tbsp Shaoxing wine
葱 3 條（切段）	• 3 spring onion (cut into short lengths)
指天椒 1 隻（切段）	• 1 bird's eye chilli (cut into short lengths)

做法

❶ 雞髀菇切厚片，菇面劃成格仔紋，備用。
❷ 熱鑊下油，下薑片及雞髀菇爆香，煎至菇軟身，下蒜粒、芝麻油、老抽及少許清水，蓋上炆煮約 5 分鐘。
❸ 開蓋，加入少許鹽及半份紹興酒調味，再蓋上繼續煮至菇熟透及收汁，下黑麻油及 1/3 份九層塔葉兜勻。
❹ 同步取另一鍋，燒熱適量麻油，先放入葱段及指天椒爆香，再加入一半九層塔炒香，倒入③，最後加入餘下九層塔葉炒勻，灒入餘下紹興酒，原鍋享用。

Method

❶ Slice mushrooms thickly. Make light crisscross incisions on each slice. Set aside.
❷ Heat wok and add oil. Fry ginger and mushrooms until fragrant. When the mushrooms are softened, add diced garlic, sesame oil, dark soy sauce and a little water. Cover the lid and cook for 5 minutes.
❸ Open the lid and add a pinch of salt and half of the Shaoxing wine. Cover the lid and keep on cooking until the mushrooms are cooked through and the juices reduce. Add black sesame oil and 1/3 of the Thai basil leaves. Toss well.
❹ In a clay pot, heat some sesame oil and stir-fry spring onion and bird's eye chilli until fragrant. Add half of the remaining Thai basil leaves. Toss until fragrant. Pour in the mushroom mixture from step 3. Add the rest of the Thai basil leaves and toss well. Drizzle with the remaining Shaoxing wine. Serve the whole pot.

必學不敗竅門

- 雞髀菇不要切得太薄，否則在炆煮時菇片會收縮。
- 在雞髀菇表面劃成格仔紋，除了美觀外還可以令菇片容易吸收味汁。
- Do not slice the king oyster mushrooms too thinly. Otherwise, they would shrink too much after braised.
- Make light crisscross incisions on sliced king oyster mushrooms. Besides looking nice, those incisions also allow the mushrooms to pick up the sauce more easily.

COOKING TIPS

素牛肉
vegetarian beef

陳皮蒸素牛肉餅

Steamed vegetarian beef patty with aged tangerine peel

材料

素牛肉 400 克
陳皮 1 瓣
檸檬葉 1 片（切幼絲）
鹽少許
胡椒粉少許
生粉 1 茶匙
葱花適量（裝飾）

Ingredients

- 400 g vegetarian beef
- 1 piece aged tangerine peel
- 1 Kaffir lime leaf (finely shredded)
- salt
- ground white pepper
- 1 tsp potato starch
- finely chopped spring onion (as garnish)

↑ 蒸後的素牛肉外觀和味道媲美真牛肉。
Vegetarian beef looks and tastes similar to real beef after steamed.

做法

1. 陳皮浸軟去瓤，一半切絲，一半切粒，水留用。
2. 素牛肉解凍後，以鹽、胡椒粉、生粉拌勻，加入陳皮粒撈勻後稍撻至起膠，平鋪在蒸碟內，用叉輕輕撥鬆，放上檸檬葉絲及陳皮絲，加入 6-8 湯匙陳皮水。
3. 大火蒸 5 分鐘，取出後灑上葱花即成。

Method

1. Soak aged tangerine peel in water until soft. Scrape off the white pith. Finely shred half of it. Dice the rest finely. Save the soaking water for later use.
2. Thaw the vegetarian beef. Sprinkle with salt, ground white pepper and potato starch. Mix well. Add the diced aged tangerine peel. Mix well and lift the beef off the bowl. Slap it back in forcefully. Repeat the lifting and slapping step until the meat slightly sticky. Transfer onto a steaming plate. Shape into a round patty. Fluff up the surface with a fork. Arrange Kaffir lime leaf and shredded aged tangerine peel on top. Drizzle with 6 to 8 tbsp of water used to soak aged tangerine peel.
3. Steam the patty over high heat for 5 minutes. Sprinkle with finely chopped spring onion at last. Serve.

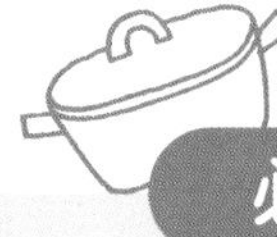

必學不敗竅門

- 素牛肉稍撻至起膠就可以，否則肉質會變硬。
- 素牛肉餅放入碟後用叉子稍稍壓平，並輕輕戳鬆，然後澆入陳皮水，讓陳皮的香氣散發素牛肉內，味道更佳。
- After mixing the ingredients into the vegetarian beef, you just need to slap it back into the mixing bowl a few times until slightly sticky. Don't overdo it. Otherwise, it may become too chewy and rubbery.
- After putting the vegetarian beef patty on a steaming plate, smooth the surface with a fork first. Then fluff up the surface slightly and pour in the water in which the aged tangerine peel was soaked. The fragrance of aged tangerine peel may then permeate through the patty and it would taste better.

OOKING TIPS

番薯

sweet potato

番薯栗子餅

Sweet potato and chestnut croquettes

材料

- 黃肉番薯 2 個
- 栗子 200 克
- 澄麵 2 湯匙
- 牛油溶液 2 湯匙
- 片糖碎 1 茶匙
- 蛋白 1 隻
- 麵包糠適量

Ingredients

- 2 yellow sweet potatoes
- 200 g chestnuts
- 2 tbsp wheat starch
- 2 tbsp melted butter
- 1 tsp crushed raw cane sugar slab
- 1 egg white
- breadcrumbs

做法

1. 栗子洗淨，焓熟後去殼、去衣，切粒備用。
2. 番薯洗淨、去皮，蒸熟後趁熱壓爛，加入澄麵、片糖碎、牛油溶液快速搓勻，加入栗子粒再搓勻成糰。
3. 取適量②搓勻成圓球狀再壓扁成餅，均勻掃蛋白後蘸上麵包糠。
4. 燒熱炸油，放入番薯栗子餅炸至金黃香脆，盛起上碟即成。

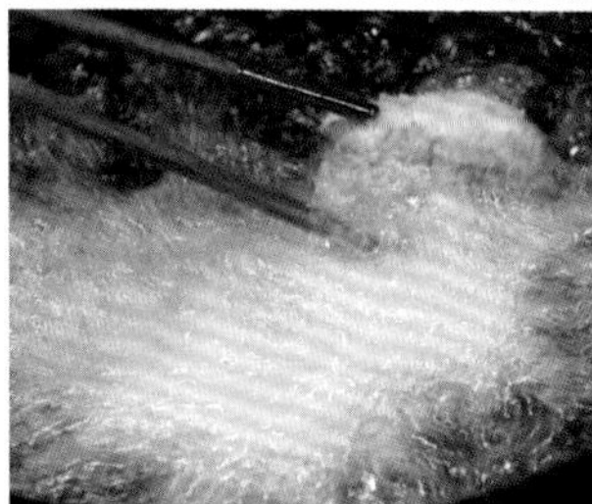

Method

1. Rinse the chestnuts. Blanch in boiling water until cooked. Shell and peel them. Dice finely.
2. Rinse and peel sweet potatoes. Steam until done and mash them while still hot. Add wheat starch, raw cane sugar slab and melted butter. Stir to mix well quickly. Put in the diced chestnuts. Mix well and knead into dough.
3. Take some of the dough from step 2. Roll it into a ball and then press to flatten it slightly. Brush egg white on the croquette. Then coat it in breadcrumbs. Repeat this step until all ingredients are used up.
4. Heat oil in a wok. Deep-fry the croquettes until golden. Save on a serving plate.

必學不敗竅門

- 番薯蒸熟後要馬上趁熱混入澄麵搓勻，否則澄麵難達至熟麵效果，炸後的番薯栗子餅會不夠酥脆。
- After steaming the sweet potatoes, stir in the wheat starch while still hot. Otherwise, the wheat starch cannot be cooked through and the croquettes won't be crispy enough after fried.

黃肉番薯及栗子。
Yellow sweet potatoes and chestnuts.

蘋果芒

Irwin mango

脆皮玉鴛鴦

Deep-fried omelette rolls with mango and honeydew

材料

- 雞蛋 3 隻
- 蘋果芒 200 克
- 蜜瓜 100 克
- 素燒肉 100 克
- 天婦羅粉 4 湯匙
- 熟蛋黃 2 個
- 泰式雞醬 1 湯匙
- 日式蛋黃醬 3 湯匙
- 素豬肉鬆少許
- 車厘茄適量（伴碟）
- 火箭菜適量（伴碟）
- 紅蘿蔔片 3 條（伴碟）

Ingredients

- 3 eggs
- 200 g Irwin mangoes
- 100 g honeydew melon
- 100 g vegetarian roast pork
- 4 tbsp tempura flour mix
- 2 yolks of hard-boiled eggs
- 1 tbsp Thai sweet chilli sauce for chicken
- 3 tbsp Japanese mayonnaise
- vegetarian pork floss
- cherry tomatoes (as garnish)
- arugula (as garnish)
- 3 strips sliced carrot (as garnish)

做法

1. 製作蛋皮：雞蛋先拂勻，加入 3 茶匙生粉水拌勻。平底鑊下油，下蛋漿以小火慢煎成直徑 12 厘米薄皮，盛起放涼備用。
2. 天婦羅粉加適量冷水拌勻成漿，備用。
3. 蘋果芒、蜜瓜起肉，與素燒肉切成相同 3 厘米條，備用。
4. 在蛋皮放上 2 條芒果條、1 條蜜瓜條及 1 條素燒肉條，包成蛋卷，以少許天婦羅漿封口，再把蛋卷均勻蘸上天婦羅漿。下油鑊炸至金黃，撈起隔油，切成一半。
5. 熟蛋黃先壓碎，加入泰式雞醬及日式蛋黃醬拌勻。
6. 把炸蛋卷排在碟上，以紅蘿蔔片圍起，放上火箭菜、車厘茄及素豬肉鬆伴碟，淋上醬汁即成。

Method

1. To make an omelette, whisk the eggs first. Add 3 tsp of potato starch thickening glaze. Mix well. Heat a pan and add oil. Pour in the whisked egg mixture. Fry over low heat into a thin omelette about 12 cm in diameter. Leave it to cool.
2. Add cold water to the tempura flour mix. Stir into a paste. Set aside.
3. Peel and de-seed mangoes and honeydew. Cut mangoes, honeydew and vegetarian roast pork into thick strips about 3 cm long.
4. Put 2 strips of mango, 1 strip of honeydew and 1 strip of vegetarian roast pork on the omelette. Roll into a cylinder. Seal the seam with some tempura batter. Then dip the whole roll into the tempura batter. Deep-fry in hot oil until golden. Drain. Cut it in half.
5. Mash the egg yolks. Add Thai sweet chilli sauce and mayonnaise. Mix well.
6. Arrange the fried omelette rolls on a serving plate. Put sliced carrot to go around the rolls. Then arrange arugula, cherry tomatoes and vegetarian pork floss on the side as garnish. Dribble with the sauce from step 5. Serve.

必學不敗竅門

- 蛋液放入生粉水拌勻才煎，蛋皮會比較煙韌不易破。
- 煎蛋皮時，如想將蛋皮煎至又圓又漂亮，宜找與蛋皮大小一樣的鑊來煎，而且油不可太多，還要用小火，並要邊煎邊轉動鑊令受熱均勻。
- Add potato starch thickening glaze to whisked eggs before frying it into an omelette. The omelette would turn out stronger in texture and is less likely to break.
- To make an omelette that looks round and pretty, find a pan of the same size as the omelette you want to make. Then try not to use too much oil and fry it over low heat. Swirl the pan continuously so that it is evenly heated.

COOKING TIPS

茄子
eggplant

魚香茄子素肉煲
Fish-scented braised eggplant and vegetarian beef in clay pot

材料

- 茄子 2 條
- 素牛肉 200 克
- 素鹹魚 2 條
- 沙茶醬 1 茶匙
- 素魚露 1 茶匙
- 紹興酒 1 湯匙
- 薑粒 2 湯匙
- 蒜粒 1 湯匙
- 紅葱頭 3 粒（拍扁）
- 葱花適量
- 紅椒粒適量

Ingredients

- 2 eggplants
- 200 g vegetarian beef
- 2 vegetarian salted-fish
- 1 tsp Sa Cha sauce
- 1 tsp vegetarian fish sauce
- 1 tbsp Shaoxing wine
- 2 tbsp diced ginger
- 1 tbsp diced garlic
- 3 shallots (crushed)
- finely chopped spring onion
- diced red chillies

做法

1. 素鹹魚解凍後切成小片，備用。
2. 茄子洗淨抹乾，連皮橫切開半，每半段再切成 2 條，再切成 4 段，備用。
3. 燒熱油鑊，放入茄子炸至香透，盛起瀝油，放入熱水內浸泡，瀝乾備用。
4. 燒熱另一鑊，加油爆香素牛肉，下少許鹽及胡椒粉調味，盛起備用。
5. 燒熱油鑊，先爆香薑粒，加入素鹹魚及紅葱粒爆香，下蒜粒及素牛肉快炒，下少許糖及沙茶醬後濳入紹興酒，再將茄子回鑊炒勻，加入約 30 毫升水煮開，倒入素魚露，最後下適量生粉芡兜勻，移入已燒熱砂鍋內，撒上葱花及紅椒粒，原鍋享用。

Method

1. Thaw the vegetarian salted-fish and cut into pieces.
2. Rinse the eggplants and wipe dry. Cut each in half lengthwise with skin on. Then cut each half into two strips. Cut each strip into four pieces.
3. Heat wok and add oil. Deep-fry eggplants until crispy and cooked through. Drain. Soak in hot water to rinse off excess oil. Drain again.
4. Heat another wok and add oil. Stir-fry vegetarian beef until fragrant. Add a pinch of salt and ground white pepper. Set aside.
5. Heat wok and add oil. Stir-fry ginger until fragrant. Add vegetarian salted fish and shallots. Stir-fry till fragrant. Add diced garlic and vegetarian beef. Toss quickly. Add a pinch of sugar and Sa Cha sauce. Drizzle with Shaoxing wine. Put the fried eggplants back in. Toss well. Pour in 30 ml of water. Bring to the boil. Add vegetarian fish sauce. Stir in potato starch thickening glaze at last and mix well. Transfer the mixture into a heated clay pot. Sprinkled with finely chopped spring onion and diced red chillies. Serve the whole pot.

必學不敗竅門

- 將茄子炸透後放入熱水內浸泡，可去除多餘油分。
- After deep-frying the eggplants, soak them in hot water to remove the grease.

冬瓜
winter melon

碧綠雙冬扒上素

Steamed winter melon globe stuffed with assorted vegetables

必學不敗竅門

- 先將冬瓜蒸至半熟，再釀入其他材料蒸至熟透，冬瓜會更透明入味。
- Just steam the winter melon bowl until half-cooked the first time around. Then stuff it with assorted vegetables and mushrooms. Then steam it until cooked through. The winter melon globe would look more transparent and taste more flavourful this way.

COOKING TIP

材料

- 冬瓜 2 斤（邊位）
- 黃耳 1-2 朵
- 榆耳 1 朵
- 靈芝菇 40 克
- 乾百合 20 克
- 馬蹄 5 粒
- 冬菇 5 朵
- 白果 10 粒
- 菜脯 3 條（切粒）
- 杞子 5 粒（浸軟）
- 白菜 8 棵（伴碟）
- 素上湯 200 毫升
- 薑（切菱形）5-6 片
- 豆瓣醬 1 茶匙
- 菇粉 1 茶匙
- 素蠔油 2 茶匙

Ingredients

- 1.2 kg winter melon (end cut)
- 1 to 2 yellow ear fungus
- 1 elm ear fungus
- 40 g Lingzhi mushrooms
- 20 g dried lily bulbs
- 5 water chestnuts
- 5 dried shiitake mushrooms
- 10 gingkoes
- 3 strips dried radish (diced)
- 5 dried goji berries (soaked in water till soft)
- 8 sprigs Bok Choy (as garnish)
- 200 ml vegetarian stock
- 5 to 6 slices ginger (cut into rhombuses)
- 1 tsp spicy bean sauce
- 1 tsp mushroom powder
- 2 tsp vegetarian oyster sauce

做法

1. 冬瓜洗淨去皮，裁成大湯碗狀，以少許鹽塗勻，蒸 20 分鐘，備用。
2. 黃耳、榆耳及冬菇浸軟（冬菇水留用）、洗淨，去蒂後以素上湯煮至入味；黃耳切細，榆耳及冬菇切片，備用。
3. 白果去殼去衣，加少許糖蒸軟，備用。
4. 乾百合洗淨、浸軟，蒸腍，備用。
5. 馬蹄洗淨去皮，切片後泡水，備用。
6. 取出冬瓜，用小刀挖空中央（瓜肉留用），放上杞子裝飾，再繼續蒸熱，備用。
7. 同步燒熱油鑊，放入薑片爆香，依次加入菜脯粒、榆耳、冬菇、黃耳、白果、靈芝菇、乾百合、馬蹄炒勻，加入豆瓣醬、適量冬菇水、菇粉、半份素蠔油炒勻，最後加入少許冬菇水生粉芡兜勻，放進瓜脯內。鋪上⑥的瓜肉，再蒸 3-4 分鐘。
8. 蒸完後取出，反扣另一碟內，伴上已焯白菜。
9. 燒熱鑊，放入冬菇水煮滾，加入餘下素蠔油及冬菇水生粉芡慢煮成玻璃芡，加入少許油拌勻，倒入瓜脯上即成。

Method

1. Rinse winter melon and peel it. Trim it so that it looks like a big bowl with rounded bottom. Rub a pinch of salt on the surface evenly. Steam for 20 minutes.
2. Soak yellow ear fungus, elm ear fungus and dried shiitake mushrooms in water till soft. (Save the water used for soaking shiitake mushrooms for later use.) Rinse all three types of mushrooms and cut off the stems. Cook them in vegetarian stock until flavourful. Then cut yellow ear fungus into smaller pieces. Slice elm ear fungus and shiitake mushrooms.
3. Shell and peel gingkoes. Add a pinch of sugar. Steam until tender.
4. Rinse lily bulbs. Soak them in water till soft. Steam until tender.
5. Rinse water chestnuts and peel them. Slice them and soak them in water. Drain right before using.
6. Cut out the centre of the winter melon with a small knife and a metal spoon. Save the flesh for later use. Garnish with goji berries. Keep steaming it.
7. Meanwhile, heat wok and add oil. Stir-fry ginger until fragrant. Then put in the followings in this particular order: diced dried radish, elm ear fungus, shiitake mushrooms, yellow ear fungus, gingkoes, Lingzhi mushrooms, lily bulbs and water chestnuts. Toss well. Add spicy bean sauce, soaking water from shiitake, mushroom powder, and half of the vegetarian oyster sauce. Toss again. Then mix some potato starch in some soaking water from shiitake mushrooms. Pour the thickening glaze in the wok. Toss until it thickens. Transfer the mixture into the steamed winter melon bowl. Cover the top with the winter melon flesh from step 6. Steam for 3 to 4 minutes.
8. Carefully remove the winter melon bowl out of the steamer or wok. Place a big serving plate on top of the winter melon bowl. Flip it upside-down so that the winter melon lies on the plate like a globe. Arrange Bok Choy around the globe after blanching them in boiling water.
9. Heat a wok. Pour in the water used to soak shiitake mushrooms. Bring to the boil. Pour in the rest of the vegetarian oyster sauce, and potato starch thickening glaze mixed with soaking water from shiitake mushrooms. Stir and cook slowly into a thin glaze. Add a dash of cooking oil. Mix well and pour it over the steamed winter melon globe. Serve.

韓國年糕

Korean rice cake

蘑菇芝士炒年糕

Fried rice cake with cheese and button mushrooms

示範短片

材料

- 韓國年糕 250 克
- 蘑菇 8 粒（切半）
- 水牛芝士 50 克
- 三文治片裝芝士 2 片
- 牛油 15 克
- 紅葱頭 1 粒（切碎）
- 蒜粒 1/2 湯匙
- 上湯 100 毫升

做法

1. 韓國年糕先以熱水浸軟，瀝乾待用。
2. 燒熱油鑊，放入年糕煎香；倒入蘑菇，略兜炒。
3. 同步燒熱另一鍋，下牛油煮溶，放入紅葱碎及蒜粒爆香，倒入上湯煮滾，放入芝士片煮成芝士汁，放入年糕和蘑菇煮至收汁，加入少許鹽及黑椒碎兜勻，上碟後放入水牛芝士，以火鎗燒至微焦融化，趁熱享用。

Ingredients

- 250 g Korean rice cake
- 8 button mushrooms (cut into halves)
- 50 g grated mozzarella cheese
- 2 slices cheddar cheese
- 15 g butter
- 1 shallot (finely chopped)
- 1/2 tbsp diced garlic
- 100 ml stock

Method

1. Soak the rice cake in hot water until soft. Drain.
2. Heat wok and add oil. Fry the rice cake until lightly browned. Put in button mushrooms and stir briefy.
3. Meanwhile, heat another pot. Put in butter and cook until it melts. Stir-fry shallot and garlic until fragrant. Add stock and bring to the boil. Put in the cheddar and cook into a sauce. Put in the rice cake and button mushrooms and cook further to reduce the liquid. Sprinkle with a pinch of salt and ground black pepper. Toss well. Transfer onto a serving plate. Sprinkle with grated mozzarella cheese. Burn the surface with a kitchen torch until the cheese melts and is lightly browned. Serve hot.

素牛肉
vegetarian beef

素擔擔麵

Vegetarian Dan Dan noodles

材料

- 上海麵餅 1 個
- 素牛肉 200 克
- 乾姬松茸 8 朵
- 鮮姬松茸 2 朵
- 青瓜 1 條

Ingredients

- 1 nest Shanghainese noodles
- 200 g vegetarian beef
- 8 dried himematsutake mushrooms
- 2 fresh himematsutake mushrooms
- 1 cucumber

擔擔麵醬

- 有粒花生醬 6 湯匙
- 麻辣醬 2 湯匙
- 麻油 2 湯匙
- 花椒油 1 湯匙
- 花生碎適量

Sauce

- 6 tbsp crunchy peanut butter
- 2 tbsp Mala sauce
- 2 tbsp sesame oil
- 1 tbsp Sichuan peppercorn oil
- crushed peanuts

做法

❶ 上海麵先汆水，盛起備用；青瓜洗淨，刨絲備用。
❷ 乾姬松茸洗淨浸軟（水留用），去掉底部後切片；鮮姬松茸先抹乾淨，切去底部，直切切片備用。
❸ 水煮沸，放入②及姬松茸水煮成素湯底。
❹ 熱鑊下油，爆香素牛肉，下少許鹽、胡椒粉調味，盛起。
❺ 製作擔擔麵醬：大碗內放進花生醬，逐少加入熱水並攪拌至合適狀，下麻辣醬、麻油，下素牛肉拌勻，最後下花椒油拌勻，待用。
❻ 把上海麵放進素湯內略煮開，撈起麵，放入碗內，加入姬松茸及素湯，放上適量擔擔麵醬，再放青瓜絲圍邊，最後撒上花生碎即成。

Method

❶ Blanch the noodles in boiling water. Drain and set aside. Rinse the cucumber. Grate it into fine strips and set aside.
❷ Soak the dried himematsutake mushrooms in water till soft. Save the soaking water for later use. Cut off the base of the stems. Slice himematsutake mushrooms. Set aside. Wipe down the fresh himematsutake mushrooms. Cut off the base of the stems. Slice lengthwise.
❸ Boil water in a pot. Put in both fresh and dried himematsutake mushrooms. Add the water used to soak the dried ones in. This is the soup base.
❹ Heat wok and add oil. Stir-fry the vegetarian beef until fragrant. Season with salt and ground white pepper. Set aside.
❺ To make the sauce, put peanut butter into a big bowl. Slowly stir in hot water to thin it out until desired consistency is achieved. Add Mala sauce and sesame oil. Put in vegetarian beef and stir well. Add Sichuan peppercorn oil and stir well.
❻ Put the noodles into the soup base from step 3. Bring to a gentle boil. Strain the noodles and transfer into a serving bowl. Pour the soup base and the himematsutake mushrooms over. Top with a scoop of the sauce mixture over. Garnish with grated cucumber on the side. Sprinkle with crushed peanuts. Serve.

必學不敗竅門

- 浸泡乾姬松茸後，須切去菇蒂底部，可避免有沙。泡菇水非常香濃，不要倒掉，可用來做素湯底。
- After soaking the dried himematsutake mushrooms, cut off the base of the stems as sand tends to accumulate there. The soaking water is very flavourful and should never be discarded down the drain. Add it to the soup base to enhance the flavour.

OOKING TIPS

芋泥

mashed taro

芋泥雪蓮子杏仁茶

Almond milk with mashed taro and honeylocust seeds

材料

- 芋頭半個
- 雪蓮子 2 湯匙
- 桃膠 4 粒
- 椰汁 80 毫升
- 南杏 300 克
- 北杏 40 克
- 蛋白 2 隻
- 冰糖 20 克

Ingredients

- 1/2 taro
- 2 tbsp honeylocust seeds
- 4 pieces peach resin
- 80 ml coconut milk
- 300 g sweet almonds
- 40 g bitter almonds
- 2 egg whites
- 20 g rock sugar

做法

❶ 雪蓮子及桃膠沖洗後用水浸泡過夜，隔掉水，各加入糖半茶匙蒸 15 分鐘，待用。

❷ 冰糖加少許水煮溶成冰糖水，待調味用。

❸ 南北杏浸泡半小時，隔掉水，放入攪拌機，加入約 800 毫升清水攪至幼滑，放入布濾袋隔渣取汁。之後將杏仁汁倒入鍋內，煮滾後改小火，邊攪拌邊煮約 20 分鐘至濃稠，加入雪蓮子及桃膠略煮熱，加入椰汁，再按個人口味加入適量冰糖水拌勻，最後加入蛋白（勿攪動，待蛋白凝固），即可熄火。

❹ 芋頭洗淨去皮，切粒後蒸約 15-20 分鐘至腍滑，放入大碗內，加入糖 1 茶匙拌勻並趁熱壓成芋蓉，待用。

❺ 上碟，取少量芋蓉壓平，置碗內，再取適量芋蓉搓成球狀放面，舀入③，趁熱享用。

Method

❶ Rinse the honeylocust seeds and peach resin. Soak them separately in water overnight. Drain. Add 1/2 tsp of sugar to each of them. Steam for 15 minutes.

❷ Cook rock sugar in a little water to make syrup.

❸ Soak both sweet and bitter almonds in water for 30 minutes. Drain. Transfer into a blender and add 800 ml of water. Puree. Strain the mixture to extract the almond milk. Pour almond milk into a pot and bring to the boil over low heat. Keep stirring while heating for about 20 minutes to reduce it. Add honeylocust seeds and peach resin. Cook until heated through. Add coconut milk and stir in rock sugar syrup according to your own taste. Pour in the egg whites at last. Do not stir them. Just let them set slowly. Turn off the heat.

❹ Rinse and peel taro. Dice it and steam for 15 to 20 minutes until tender. Transfer into a big bowl. Add 1 tsp of sugar and mix well. Mash it.

❺ Spread some of the mashed taro on the bottom of a bowl. Roll some mashed taro into a ball and arrange on top. Pour the almond milk sweet soup from step 3 over. Serve hot.

↑ 雪蓮子 honeylocust seeds

↑ 桃膠 peach resin

必學不敗竅門

- 雪蓮子和桃膠先用清水隔夜浸泡，煮前加適量糖蒸片刻，令雪蓮子和桃膠更入味。
- After soaking the honeylocust seeds and peach resin in water overnight, add sugar and steam them. This step would let sweetness permeate through the honeylocust seeds and peach resin.

OOKING TIPS

雪梨

Chinese pear

陳皮話梅燉雪梨

Double-steamed Chinese pear with aged tangerine peel and dried plum

材料

- 雪梨 1 個
- 話梅 1 粒
- 陳皮 1 小瓣

Ingredients

- 1 Chinese pear
- 1 dried liquorice plum
- 1 small piece aged tangerine peel

做法

1. 陳皮浸軟（水留用），備用。
2. 雪梨洗淨，切去頂部小部分，去皮去芯，挖空中間，須挖淨核芯，但勿挖穿底部。
3. 將梨放入燉盅內，放入話梅及陳皮，倒入陳皮水，再注水入梨旁至半滿，蓋好盅蓋燉 1 小時即成。

Method

1. Soak aged tangerine peel in water until soft. Save the soaking water for later use.
2. Rinse the pear and cut off the top. Peel and core it. Scoop out the centre and make sure the seeds and core are removed thoroughly. Do not cut through the bottom.
3. Put the pear into a double-steaming pot. Put in the dried plum and aged tangerine peel. Pour in the water used to soak the aged tangerine peel. Add fresh water up to half full. Cover the lid and steam in a steamer for 1 hour. Serve.

必學不敗竅門

- 嗜甜者可加入一小粒冰糖同燉，味道會更甜美。
- Those who prefer a sweeter taste may as add a small cube of rock sugar before double-steaming.

COOKING TIP

木瓜

papaya

木瓜鮮奶燉桃膠

Double-steamed milk and peach resin in whole papaya

材料

- 木瓜 1 個
- 鮮奶 150 毫升（視木瓜大小而增減）
- 桃膠 3 粒
- 鮮百合少許
- 冰糖碎 1-2 茶匙

Ingredients

- 1 papaya
- 150 ml milk (use more or less according to the size of papaya)
- 3 pieces peach resin
- fresh lily bulbs
- 1 to 2 tsp crushed rock sugar

做法

1. 桃膠洗淨、浸泡過夜，倒掉水後加入少許糖蒸軟，備用。
2. 木瓜橫切開頂 1/5，挖掉瓜瓤及籽。
3. 將桃膠、鮮百合及冰糖碎放入木瓜內，倒入鮮奶，蓋上木瓜頂，封上耐熱保鮮紙，燉 30 分鐘即成。

Method

1. Rinse the peach resin and soak it in water overnight. Drain. Add some sugar and mix well. Steam until soft.
2. Cut off 1/5 of the papaya along the length. Scoop out and discard the seeds.
3. Put the peach resin, fresh lily bulbs and crushed rock sugar into the papaya. Pour in milk. Cover with the cut-off top. Wrap in microwave cling film and steam it for 30 minutes. Serve.

必學不敗竅門

- 宜選用半生熟的木瓜，因為太熟的木瓜容易在燉時溶掉。
- For this recipe, pick a half-ripe papaya. Fully ripe papaya would turn mushy and break down in the steaming process.

OOKING TIPS

著者
蕭秀香（三姐）· 江美儀

策劃
謝妙華

責任編輯
簡詠怡、譚麗琴

食譜撰寫
Jackie、Carmen

翻譯
Wendell J. Leers

封面設計
陳重衡

裝幀設計
鍾啟善

排版
辛紅梅、何秋雲、劉葉青

相片提供
《女人必學 100 道菜》製作組

出版者
萬里機構出版有限公司
香港北角英皇道 499 號北角工業大廈 20 樓
電話：(852) 2564 7511
傳真：(852) 2565 5539
電郵：info@wanlibk.com
網址：http://www.wanlibk.com
http://www.facebook.com/wanlibk

發行者
香港聯合書刊物流有限公司
香港荃灣德士古道 220-248 號荃灣工業中心 16 樓
電話：(852) 2150 2100
傳真：(852) 2407 3062
網址：http://www.suplogistics.com.hk

承印者
中華商務彩色印刷有限公司
香港新界大埔汀麗路 36 號

規格
特 16 開（240mm x 170mm）

出版日期
二〇二〇年九月第一次印刷
二〇二四年一月第六次印刷

Published in Hong Kong, China.
Printed in China.
ISBN 978-962-14-7282-3

本書由電視廣播有限公司授權出版